Contents

RFF REPORT

ASSIGNING LIABILITY FOR SUPERFUND CLEANUPS
An Analysis of Policy Options

KATHERINE N. PROBST
PAUL R. PORTNEY

RESOURCES FOR THE FUTURE
WASHINGTON, D.C.
JUNE 1992

Resources for the Future publishes RFF Reports to facilitate the wide and public distribution of their contents. Reports have not been subject to the same formal review and editing procedures of books from Resources for the Future. The views expressed should be attributed to the authors alone. They do not represent the positions of Resources for the Future as an institution.

Library of Congress Catalog Card Number 92-60830

Published by Resources for the Future
1616 P Street, NW; Washington, DC 20036-1400

Books from Resources for the Future are distributed worldwide by The Johns Hopkins University Press.

Foreword

In March of 1987, Resources for the Future (RFF) announced the creation of its Center for Risk Management. The purpose of the Center is to stimulate discussion about creative ways to manage a wide variety of risks to human health, public safety, and the environment. The Center conducts research on critical risk and policy issues and places particular emphasis on outreach and educational functions. It aims to involve in its work all of the groups that have an interest in possible new approaches to addressing environmental policy problems.

Appropriately, then, this report had its genesis in several meetings that Center researchers held in 1989 with representatives of the U.S. Environmental Protection Agency, business groups, environmentalists, academics, and others. The purpose of those meetings was to help identify new areas of emphasis for the Center, which was completing its first round of projects, initiated in 1987. There was nearly unanimous agreement among those consulted that the Center should mount a project related to the ambitious and increasingly controversial Superfund program. Center fellow Katherine Probst and acting Center director and RFF vice president Paul Portney set out to undertake such a project. The fruits of their efforts are contained herein.

It is worth saying a word or two about how the authors proceeded with this project. Their first step after deciding to focus on Superfund liability issues was to talk with a broad range of parties interested in the Superfund program. The meetings that they held both informed "stakeholders" about the likely focus of the RFF study and also provided valuable direction to Probst and Portney as they set out to do their research and policy analysis. Following the completion of a draft report, they solicited comments from more than 40 experts in government, academe, environmental organizations, the business and legal communities, and others. In a sense, then, this report has been subject to both "notice" and "comment," to use rule-making parlance. Of course, this report does not represent a consensus of the views solicited—the opinions expressed are those of the authors.

This report is meant to help encourage and enlighten discussion about the Superfund program and about the ways it might be made more efficient, effective, and fair. If the report has that effect—and I believe it will —it will make exactly the kind of contribution RFF intended when it established the Center for Risk Management.

Robert W. Fri
President
Resources for the Future

May 1992

Acknowledgments

Many people have given freely of their time over the past 18 months to help make this a better report. Because they are so numerous, we cannot thank them all by name.

We are indebted to those with whom we met in the early months of the project who helped us define the options and the criteria against which they are measured. After the basic framework for the study was developed, more than fifty people suggested ways in which we could improve the analytical framework of this report. Throughout our work, many staff people at the U.S. Environmental Protection Agency (EPA) helped us by providing reliable and up-to-date information about Superfund and answering questions about the program. In addition, we worked closely with analysts at the U.S. General Accounting Office and the Congressional Budget Office who shared their insights about the Superfund program, often gleaned from their own studies. Finally, more than forty people—from EPA, environmental groups, private parties, insurers, and other organizations—reviewed a previous draft of this report and gave us helpful and constructive comments. Of course, the views expressed here are our own, but we wish to thank all those who gave so generously of their time to help make this a better analysis.

We are also grateful to Matthew Farrelly, Jonathan Taylor, and Christine Wnuk for their research assistance; to John Mankin for typing many drafts during the entire project period; to Chris Mendes and Anne Jarrett for helping with many of the final production details; to Dorothy Sawicki for her expert editorial assistance; to Marilyn Voigt for help and support throughout the preparation of the report; and to Ann Checkley for efficiently managing its distribution.

Finally, we are indebted to those that provide financial support for RFF's Center for Risk Management. Since no project-specific funds were solicited for the preparation of this report, the general support of the Center made this report possible.

Executive Summary

The next reauthorization provides an opportunity to debate in an open and comprehensive fashion whether fundamental changes should be made in any facets of the Comprehensive Environmental Response, Compensation, and Liability Act (better known as Superfund), including its liability standards. That opportunity should not be missed.

Superfund has been a controversial statute almost from the start. Although more than 2,700 emergency removals of hazardous materials have taken place under the law, the implementation of the long-term cleanup (or remedial) program has been the object of considerable dissatisfaction. Environmentalists decry the slow pace of cleanup and are disturbed by what they see as less-than-permanent remedies being put in place. Potentially responsible parties (PRPs) resent the retroactive imposition of liability and the sometimes-significant legal fees and other "transactions costs" accompanying Superfund litigation. Municipalities and banks and other lending institutions object to being held liable for cleanup costs they argue they were never intended to pay; and insurers are concerned about potential losses under general liability policies they wrote years ago.

Superfund is likely to be subjected to very careful scrutiny in the reauthorization debate that will begin in earnest in the 103d Congress in 1993. This makes it a most propitious time for an independent, nonpartisan examination of the law. Our purpose here is to undertake such an analysis, focused on how liability is assigned for cleanups. Indeed, the explicit purpose of this report is to stimulate informed debate both about the accomplishments under the current liability provisions of Superfund and about the advantages and disadvantages associated with possible changes.

In this report we evaluate the current Superfund financing scheme and four alternative liability approaches and their likely effect on current and future sites on the U.S. Environmental Protection Agency's (EPA's) National Priorities List (NPL). For each of the four alternative approaches we consider, we assume that increases in the Hazardous Substance Response Fund (the Trust Fund) needed to pay for additional EPA cleanups come from an increase in the corporate environmental tax. In actuality, of course, increased revenues for the Trust Fund could come from a variety of sources. The options are summarized below.

Five Policy Options

Option 1. Status Quo (The Current Superfund Program). Potentially responsible parties (PRPs) include the past and present owners and/or operators of a site, those who arranged for the transportation of hazardous substances to the site, and those who arranged for the treatment or disposal of the substances. These PRPs are subject to retroactive, strict, and joint and several liability. Under the current law, the Trust Fund generates approximately $1.5 billion annually. The Trust Fund is financed by petroleum excise taxes, a chemical feedstock tax, a corporate environmental tax, general revenues, and other sources. These funds are used by EPA to conduct site studies and cleanups itself and to finance the overall Superfund program, which involves settlement negotiations and enforcement actions to get PRPs to pay for site activities, as well as other aspects of the Superfund program.

Option 2. Expanded Mixed Funding for Orphan Shares. Under this option, the Trust Fund would be increased to cover the shares of insolvent and recalcitrant parties at all current and future NPL sites. The liability standards in Superfund would remain unchanged, but, as a matter of policy, EPA would "mixed fund" with PRPs to cover orphan shares. That is, EPA would pay for the shares of insolvent and recalcitrant parties and would use the law's powerful enforcement tools to seek to recover costs from recalcitrant parties. For the purposes of evaluating the financial implications of this option, we assume that just under 75 percent of current NPL sites are multiparty sites with orphan shares and that the orphan share at a site averages 10 percent of total costs. While this option would release PRPs from liability at some future NPL sites, we do not estimate the financial implications of these additional NPL sites.

Option 3. Liability Release for All Closed Co-disposal Sites. Under this approach, any sites that accepted both municipal solid waste and industrial waste and that were closed at the time Superfund is amended would be cleaned up by EPA using Trust Fund monies. Thus, this option would apply to both current and future NPL sites. All PRPs at these sites would be released from Superfund liability. Based on our analysis of existing data, we assume that there are approximately 250 such sites on the current NPL. While some future NPL sites will probably meet the criteria established above for a liability release, we do not estimate the financial implications for these future NPL sites.

Option 4. Liability Release for All Pre-1981 Sites. All PRPs at all sites where the waste disposal operation closed before January 1, 1981, would be released from Superfund liability under this option. Using monies from an augmented Trust Fund, EPA would be responsible for cleaning up these sites. This release would apply to sites that meet this criterion that are added to the NPL in the future. We estimate that there are 580 sites on the current NPL that meet this criterion.

Option 5. Liability Release for Current NPL Sites. Under Option 5, all PRPs at all sites on the NPL on the date Superfund is reauthorized would be released from responsibility for cleanup of these sites. EPA would take over the responsibility and pay for cleanup at these sites using an expanded Trust Fund. This release from liability would apply only to sites on the NPL when the statute is amended; it would *not* apply to any sites added to the NPL after the statute is amended.

Evaluative Criteria

There are dozens of standards against which the options considered could be evaluated. The dissatisfaction with the present statute revolves primarily around the slow pace of cleanups, the transactions costs it generates, and its perceived unfairness. For this reason, these are logical criteria to use in evaluating possible alternative liability standards. In addition, the current liability provisions are thought by some to induce cleanups at sites not yet on the NPL and also to provide incentives for careful handling of hazardous substances that go beyond those provided by regulations under the Resource Conservation and Recovery Act (RCRA)—which pertains to a smaller universe of hazardous substances than Superfund does—and by other statutes. To reflect these two concerns, we also include voluntary cleanups at non-NPL sites and due care in future waste management practices as evaluative criteria. Finally, we are concerned with the financial implications of shifting cleanup responsibility from PRPs to the Trust Fund.

To summarize, then, our criteria are as follows:

- Speed of cleanup
- Transactions costs
- Voluntary cleanups at non-NPL sites
- Due care in future waste management practices
- Fairness
- Financial implications

We do not include the degree of protection afforded by cleanups as a criterion because we assume throughout our analysis that the standards that cleanups must meet stay the same under each of our options.

Conclusions

Our analysis of the five policy options identifies strengths and weaknesses of each. Based on this analysis, we reach a number of general conclusions.

The site cleanup process could be expedited by changing Superfund's liability standards. It could also be expedited under the current system.

Some cleanup delay results from the need to negotiate with PRPs to get them to agree to conduct site studies and cleanup. Releasing PRPs from some or all liability at some number of NPL sites (as in Options 2 through 5) should reduce these delays. While we cannot be sure how much cleanup would be accelerated, the cleanup process could be expedited considerably at sites where many parties are involved and litigation is rampant. Just as clearly, however, the remedial process could be accelerated under the existing liability standards, as EPA has recently proposed. Thus, while it is unarguable that changing the liability standards would speed cleanup, it is less clear whether that is the most promising way to accelerate the cleanup of sites on the National Priorities List.

Transactions costs could be reduced through modification of Superfund liability.

Eliminating the need for EPA to reach agreement on cleanups with PRPs by releasing them from liability (under Options 3 through 5) would reduce transactions costs, but there is presently too little information available on the current magnitude of transactions costs to know just how large the cost savings would be. Other alternative liability approaches, such as Option 2, have the potential to reduce transactions costs resulting from litigation between EPA and PRPs, and among PRPs, but would do little to reduce insurance coverage litigation. Options 3 through 5, however, would release all PRPs from liability at certain sites, and thus might eliminate all transactions costs at these sites. We *speculate* (and it is no more than this) that the transactions costs savings would be in the range of $2 billion to $8 billion, albeit spread out over 10 years. None of the options we evaluate would eliminate *all* transactions costs at all NPL sites.

Relaxing Superfund's liability standards would have some adverse effects.

Any alternative that eliminates Superfund liability for a subset of sites could diminish—if not eliminate—the current incentives PRPs face both to clean up sites not on the NPL and also to carefully handle hazardous substances not regulated under other statutes. The effect of any change in liability on non-NPL cleanups is straightforward. Any release from liability for a specific category of sites (as in Options 3 and 4) would greatly diminish the incentive to clean up similar sites not on the NPL. It is much more difficult to determine how a change in liability would affect incentives for the careful management of hazardous substances. Logically, eliminating the retroactive aspect of Superfund liability, as in Option 4, should have no effect on prospective waste management practices. Nevertheless, the imposition of retroactive liability in 1980 at least initially jolted everyone dealing with hazardous substances into recognizing these substances' potentially harmful effects. We cannot discount the possibility that abolition of retroactive liability would lead to backsliding by at least some PRPs in their future waste management practices.

Any of the modifications to the present liability standards will create at least some new inequities, even as they ameliorate others.

Any overall evaluation of the options we consider on grounds of fairness depends on which concept of fairness one prefers. We have identified three possible interpretations of fairness: that those who created the problem should bear the costs of remediating it (the "polluter-pays" principle); that the liability standards should treat as equals all PRPs who handled hazardous substances in a similar fashion; and, finally, that rule changes be made prospectively, not retroactively. Each of the options we consider does well against one or two of these three possible interpretations of fairness; none does well against all three. For example, eliminating retroactive liability would address one possible inequity created by Superfund. But using the Trust Fund to clean up a site owned and operated by one company throughout the life of the site would be a clear violation of the polluter-pays concept. Similar conflicts arise with respect to each option we consider.

Much better data are needed to assess the financial implications of the present liability standards as well as any proposed alternatives to them.

The United States does a very poor job of keeping track of any expenditures related to Superfund other than those made by EPA. This makes it extremely difficult to estimate the full financial and economic implications of any liability option, including the current program. There is a great deal of uncertainty about each of the building blocks on which our cost estimates are based: the average cost of a cleanup, the number of sites likely to be affected by each of the options, the time horizon over which cleanup will take place, the share of total expenditures at a site devoted to legal fees, and so on. Any serious consideration of the future funding needs for Superfund—even absent any changes to the law— must be based on a better assessment of the costs of the current program to both EPA and PRPs.

This report is silent on two important issues relating to changes in Superfund's liability standards and taxing scheme: the effect of any change in liability on the kinds of remedies selected and the ability of EPA to effectively manage a large number of site cleanups at one time. These issues must be borne in mind if serious consideration is given to changing Superfund's liability scheme.

We assume throughout this report that remedy selection at NPL sites would not be affected by a change in the origin of the cleanup funds. We could see no other way to proceed. Nevertheless, if government financing of remedial actions through the Trust Fund leads to either gold-plated or shoddy remedies, this must be factored into an evaluation of possible legislative changes. Similarly, in our analysis we assume that EPA could manage effectively even a great expansion in the number of sites cleaned up using the (augmented) Trust Fund. This assumption would also have to receive careful scrutiny if any serious consideration is given to modifications in the present liability standards.

For all the attention paid to the costs of the Superfund program, they may pale in comparison with the costs of cleaning up problems at federal facilities (particularly those of the Department of Energy) and also in comparison with the costs of "corrective action" cleanups that will be required under RCRA.

Although Superfund has garnered much attention, there are other cleanup programs— notably those at federal facilities and those that will be required under RCRA—where the cleanup costs could dwarf those of Superfund. These programs have not received as much attention as Superfund for a variety of reasons. The transactions costs associated with these cleanup programs may be smaller (since they are more akin to single-party than to the more heavily litigated multiparty sites under Superfund); nevertheless, the fundamental public policy questions about where we are spending our money and what we are getting for it could not be more pertinent. We hope these other important cleanup programs come under the same bright light now being directed at Superfund.

1
Introduction

The Comprehensive Environmental Response, Compensation, and Liability Act (better known as Superfund or CERCLA) was rushed into being in the early morning hours of December 3, 1980. It was enacted by a lame-duck Congress and signed by a lame-duck president, both apprehensive about the likely environmental policies of President-elect Ronald Reagan. Perhaps befitting this frenzied origin, Superfund has been a controversial statute almost from the start.

While more than 2,700 emergency removals of hazardous materials have taken place under Superfund,[1] the implementation of the long-term cleanup (or remedial) program has been the object of considerable dissatisfaction. Environmentalists and community groups frequently decry the slow pace of cleanups under the law, and are also disturbed by what they view as the less-than-permanent remedies that are being put in place at a number of sites. Manufacturers and other entities, public or private, that are potentially responsible parties (PRPs) under the law likewise have strong complaints. They resent the retroactive imposition of liability for practices that were legal when they were carried out and the sometimes-significant legal fees and other "transactions costs" accompanying Superfund litigation and settlement negotiations. Moreover, they often bristle at having stepped forward to shoulder their share of site cleanup costs only to see other PRPs do better for themselves by "lying in the weeds"—that is, waiting to see if either the U.S. Environmental Protection Agency (EPA) or participating PRPs decide to sue them over their involvement in a particular site. In addition, municipalities and banks and other lending institutions object to being held liable for cleanup costs that they argue they were never intended to bear under Superfund. Finally, insurers are concerned about the losses they will incur if courts hold that the general liability policies they wrote years ago apply to site cleanups. While everyone seems upset, there is distressingly little agreement about what should be done to alleviate these concerns.

For a number of reasons, Superfund is likely to be subjected to very careful scrutiny in the reauthorization debate that will begin in earnest in the 103d Congress in 1993. First,

---

[1] See testimony of William K. Reilly, Administrator, U.S. Environmental Protection Agency, before the Subcommittee on Superfund, Ocean and Water Protection of the Senate Committee on Environment and Public Works, April 8, 1992, 102d Cong., 2d Sess.

a spate of recent studies has begun to call attention to the current and likely future expenditures that Superfund cleanups will necessitate.[2] Even the most optimistic forecast of cleanup expenditures is on the order of $60 billion to $90 billion over the next several decades; this has helped draw attention to the economic significance of the Superfund program.

Second, concern has been expressed by some that the quantified risks to health and the environment posed by many Superfund sites may be small relative to those arising from other environmental problems. Importantly, this view is not confined exclusively to those in the business community. In fact, the most influential of the reports calling into question the seriousness of hazardous waste site risks have emanated from EPA itself.[3] Coupled with the growing awareness of both site remediation costs and the limitations of cleanup technologies for certain kinds of contamination, this lack of consensus on site cleanup goals has raised serious questions about the cost-effectiveness—or "bang-for-the-buck"—of Superfund cleanups.

Third, and somewhat at variance with this last point, is the growing recognition that we know embarrassingly little about the actual magnitude of risks to human health and the environment at most Superfund sites. A committee of the National Academy of Sciences recently completed the most thorough review to date of information about these health risks; the committee concluded that existing data provide no basis for determining whether or not Superfund sites pose serious risks to health.[4] That this is so more than a decade after Superfund was passed reinforces the case for a thorough analysis of the program.

Fourth, regardless of how expensive it might be to clean up the sites on the National Priorities List (NPL)—which is EPA's list of hazardous waste sites determined to be in greatest need of remediation—the public wants the job done. Several recent public opinion polls dealing with environmental issues have come to the same conclusion: the public cares more about hazardous waste sites than it does about any other single environmental problem.[5] This fact alone is sufficient to ensure that elected officials at all levels of government will continue to pay close attention to progress (or lack thereof) in site cleanups.

Fifth, in 1990 Congress elected to enact a simple reauthorization of Superfund as part of the Omnibus Budget Reconciliation Act of 1990, without reopening the statute to substantive changes. When the 103d Congress convenes in 1993, it will have been seven

---

[2] Robert W. Hahn, "Reshaping Environmental Policy: The Test Case of Hazardous Waste," *The American Enterprise* vol. 2, no. 3 (1991) pp. 73-80; E. William Colglazier, Tracy Cox, and Kim Davis, *Estimating Resource Requirements for NPL Sites* (Knoxville, Tenn., University of Tennessee, Waste Management Research and Education Institute, 1991); Paul R. Portney, "The Economics of Hazardous Waste Regulation," in *U.S. Waste Management Policies: Impact on Economic Growth and Investment Strategies* (Washington, D.C., American Council for Capital Formation Center for Policy Research, forthcoming 1992).

[3] U.S. Environmental Protection Agency, *Unfinished Business: A Comparative Assessment of Environmental Problems* (Washington, D.C., EPA, 1987); U.S. Environmental Protection Agency, Science Advisory Board, *Reducing Risk: Setting Priorities and Strategies for Environmental Protection* (Washington, D.C., EPA, 1990).

[4] National Research Council, Committee on Environmental Epidemiology, *Environmental Epidemiology: Public Health and Hazardous Wastes* (Washington, D.C., National Academy Press, 1991).

[5] Environmental Opinion Study, Inc., *1991 National Survey* (Washington, D.C., EOS, June 1991); Roper Organization, "The Environment: Public Attitudes and Individual Behavior," commissioned by S. C. Johnson & Son, New York, N.Y. (July 1990).

years since a number of substantive issues in Superfund were last addressed legislatively (that is, in 1986 with the passage of the Superfund Amendments and Reauthorization Act, known as SARA). For this reason, too, then, the statute is almost certain to be scrutinized and debated openly in the next several years.

All in all, it is a most propitious time for an independent, nonpartisan examination of Superfund. Our purpose here is to undertake such an analysis focused on how liability is assigned for cleanups.

The multitudinous issues wrapped up in Superfund seem ultimately to boil down to two basic questions: (1) What is the appropriate degree of cleanup at Superfund sites (the "How clean is clean" question)? and (2) What is the best means of financing cleanups (that is, Who should pay the cleanup costs)? Certainly the two questions are closely linked. For example, if it were determined that only perfunctory remedial efforts were appropriate at most Superfund sites, it would be much easier to find parties willing to step forward and participate in site cleanups. Conversely, a policy decision to excavate all contaminated soils at Superfund sites and to treat or incinerate these soils off site would greatly increase the cost of cleanups. This in turn would affect the success of any liability-based financing option. This report, however, concentrates only on the second of these questions—Who should be liable for cleanup costs?

Few issues in U.S. environmental policy today are more worthy of careful analysis and informed debate than the appropriate extent of cleanup at the sites on the NPL. We may, in fact, address this question ourselves in future work. However, two major reasons led us to concentrate in the present report on liability issues under Superfund. First, the current statute expresses a strong preference for "permanent" remedies (that is, fairly stringent cleanup standards) at Superfund sites. That being the case, it is worth knowing whether changes in the current liability scheme in Superfund will speed the attainment of these standards and advance other possible objectives of the law that are widely regarded as desirable.

The second reason for our concentration on liability issues under Superfund is that when we began our work more than a year ago, it was clear that the current state of quantitative risk assessment would make virtually impossible anything other than sweeping generalizations about the marginal risk reductions and marginal costs associated with alternative cleanup strategies at the sites on the NPL. For this reason, too, it seemed wise to focus on a more manageable assignment.

Plan of the Report

Our approach in this report is straightforward. Chapter 2, "Policy Options and Evaluative Criteria," first provides a brief description of the current liability and taxing provisions in Superfund; the current program, called the "status quo," is referred to in this report as Option 1. Chapter 2 then identifies four possible alternative approaches—Options 2 through 5. These alternatives range from a relatively slight modification of the current system to wholesale abandonment of the current liability approach for all sites on the NPL at the date of enactment of an amended Superfund statute. Following the identification and description of Options 1 through 5, the chapter discusses the criteria against which we later evaluate both the current program and the four alternative approaches.

Chapter 3 presents our analysis of the five policy options. Each of the options is evaluated individually against the six criteria identified in chapter 2. Option 1 forms the baseline against which the alternative approaches are evaluated.

Our "methodology" is quite simple. We began by collecting as much reliable information as we could about the current Superfund program. In the course of conducting the project, we integrated a wealth of EPA data to develop a Superfund database that provided a very helpful picture of the current universe of sites on the NPL, the state of cleanup at those sites, and other site-specific data. Wherever possible in the analysis in chapter 3, we rely on these data to support the inferences we draw about the current program and alternative approaches. It will become obvious, however, that data are in short supply, even where important questions call for answers.

In addition to this data collection and analysis, we talked formally or informally with well over a hundred participants in the Superfund policy debate. These included policymakers and program analysts at EPA; administration officials and legislative staffers; members of both national and local environmental interest groups; representatives from the business community—including manufacturers, insurers, and waste handlers; and lawyers and academics with a strong interest in the Superfund program. These discussions informed us and helped shape our views on both the current Superfund program and the effects of possible changes in its liability standards. Nevertheless, this report is not a consensus document; instead, it reflects our own views and interpretations.

Ultimately, of course, we made use not only of the information we collected but of our own analytical skills and intuition in our analysis. We wish to state clearly here that the final evaluations, presented in chapter 3, are based on our judgment and ours alone. We expect that others not only will differ with our assessments, but also may champion other alternatives or prefer a different set of evaluative criteria. This bothers us not at all. **Indeed, the explicit purpose of this report is to stimulate informed debate both about the accomplishments under the current liability provisions of Superfund and about the advantages and disadvantages associated with possible changes.**

In the final chapter, we step back from the detailed analysis of options in chapter 3 to make several general observations concerning the existing liability standards in Superfund and the alternatives to them that we have considered.

This is an appropriate point at which to raise a caution: those taking up this report eager to find the "winning option" will be disappointed. One of the conclusions we have drawn over the past 18 months is that there is no easy or obvious change that can be made in Superfund's liability provisions that will improve the working of the law in every important respect. Each of the possible changes we consider has the potential to improve the Superfund program in one or perhaps several dimensions, while at the same time creating other potential difficulties. Whether any of the options we consider is preferable to the current law depends squarely on the weights one attaches to the evaluative criteria used.

As noted above, we do not deal in this report with the important question of cleanup standards at Superfund sites. Several other issues pertinent to the Superfund debate are also not addressed. For instance, we do not deal explicitly with the question of lender liability. Although banks and other financial institutions are vitally interested in this subject, we are more concerned with the assignment of initial liability to traditional PRPs and the litigation to which that gives rise. Nevertheless, some of the analysis presented here bears at least indirectly on this question. Specifically, several of the alternatives considered would exempt certain types of responsible parties or certain types of sites from cleanup liability. To the extent that this lessens the likelihood that lending organizations will be drawn into the Superfund "net," it may be of interest to them, but we do not offer our judgment as to the legitimacy or wisdom of extending liability to such organizations.

Nor do we discuss the applicability of general liability insurance policies to Superfund cleanup costs. This is largely an issue of state law (albeit one with very important economic ramifications) that falls outside our areas of expertise. As with the lender liability question, however, the issues we examine are of considerable relevance to insurers. If their policy-holders are relieved of liability for certain cleanup costs, insurers are left better off as well. When the alternative approaches analyzed in chapter 3 would affect the likelihood of insurance coverage litigation, we so note.

Our work does not address the role of government agencies other than EPA in Superfund activities. Specifically, we do not discuss the role of the U.S. Department of Justice in the Superfund enforcement process, or the costs of cleaning up "federal facilities" (that is, those sites identified as warranting cleanup that are owned or operated by federal agencies such as the U.S. Department of Defense or of Energy). We do not analyze the latter because funding for cleanups at federal facility sites comes directly from general revenues. Nor do we say much about one other group of key players in the Superfund program—the states. Although the states play a crucial role in site cleanups—not only may they implement many Superfund activities in lieu of EPA, but many states also have their own Superfund programs—we do not address the state-EPA relationship. However, any changes in liability standards in the federal Superfund law could have major effects on state governments by creating pressure for similar changes in state laws and/or by affecting the number of cleanups for which state contributions are required.

Finally, we do not address the financial implications of natural resource damage litigation in this report; we do not extend the analysis to include the costs of restoring damaged resources, or to payments that may be required to compensate for the lost use of the resources during the time they were damaged. We confine our attention to who is liable for the costs of remediating Superfund sites.

2

Policy Options and Evaluative Criteria

Despite the dissatisfaction with Superfund noted in chapter 1, surprisingly few concrete proposals for changes in its liability and taxing mechanisms have been put forward. (Two exceptions are noted below.) Had a variety of such proposals been advanced, our analysis could have consisted of identifying appropriate evaluative criteria and analyzing each proposal against them. Given the lack of well-defined options, however, we elected to create a variety of possible approaches ourselves.

Of the two significant alternatives that have been put forward, one was proposed by the American International Group (AIG), the largest underwriter of commercial and industrial insurance in the United States. AIG suggested in March 1989 that retroactive liability under Superfund for "old sites" be replaced with a dedicated National Environmental Trust Fund (NETF). This proposal did not make clear exactly how "old sites" would be defined or how many sites would be affected. Revenues for NETF, according to the proposal, were to come from a 2 percent surcharge on commercial and industrial insurance premiums in the United States and an equivalent charge on firms that self-insured. Subsequent to making its original proposal, AIG (which was joined in promoting the NETF concept in late 1989 by Crum and Forster Insurance Companies and Fireman's Fund Insurance Companies) has indicated flexibility with respect to the financing mechanism that it could support.

A second proposal, sometimes referred to as a municipal "carve out," is motivated by concern over so-called contribution lawsuits brought by corporate PRPs against municipalities. Under this proposal—embodied in legislation introduced by Congressman Robert G. Torricelli (D-N.J.) (H.R. 3026) and Senator Frank R. Lautenberg (D-N.J.) (S. 1557)— a municipality or other entity would be exempt from cleanup liability at a site to which it had contributed no wastes other than household garbage or sewage sludge. The proposed legislation would release certain waste generators and transporters from liability but would retain Superfund liability for the owners and/or operators of these facilities. (In this report, we consider one option [see the discussion of Option 3 in the next section] that addresses all closed landfills at which both municipal solid wastes and industrial wastes were disposed of—referred to here as co-disposal sites—but the option goes beyond the legislative proposals discussed above in that it would release *all* PRPs at these facilities from Superfund liability.)

In creating possible alternative approaches ourselves, we had to make two decisions. First, for each alternative we needed to determine the respects in which the current liability provisions under Superfund would change. Since the current law casts a very broad liability net, the new options we identify all involve some relaxation of the current liability standards. Under one alternative (see the discussion of Option 2 in the next section), no formal change would be made in the liability standards per se, but EPA would be more aggressive in funding so-called "orphan shares" at sites. (We define orphan shares as those of insolvent or recalcitrant parties, as explained later in this chapter.) Each of the other alternatives involves some significant relaxation in the current liability standards.

The second issue we had to address in creating each alternative approach was how cleanup would be paid for at sites where PRPs would no longer be held liable. In other words, where would the additional monies come from? There are many possibilities. For instance, the funds could come from general revenues; from a national sales or value-added tax, a tax on industrial and/or municipal waste generation, or an increase in one or more of the taxes used to generate the existing Hazardous Substance Response Fund (hereafter referred to as the Trust Fund), which is the source of revenues to pay for EPA's Superfund activities; or from any number of other possibilities (including reduced spending on one or more current federal programs).

Ideally, of course, there would be a tax on the generation of those hazardous substances that are capable of creating the Superfund sites of the future; such a "waste end" tax would be an excellent way to raise the revenues needed for remediating past problems. In addition to generating revenues to clean up Superfund sites, it could act as a powerful incentive for waste minimization or for treatment of wastes prior to disposal. For a variety of reasons, however, Congress rejected such an approach in SARA. Despite its appeal, a tax on hazardous waste generation seems unlikely to be resurrected in 1993.

Rather, it seems more likely to us that any additional revenues that might be required for the Trust Fund would come from a broad-based tax on the corporate sector. Since the current corporate environmental tax is just such a tax, for each option we consider that releases some PRPs from liability, we estimate how much the corporate environmental tax would need to be increased to cover the added costs to EPA for cleanups. We do not mean to suggest that additional Trust Fund revenues *should* come from this tax, but because it is a likely source of added revenues, we analyze it here. Readers are encouraged to envision and advance other taxing mechanisms.

In this report we evaluate the current Superfund financing program and four alternative approaches, which are described in the next section. We refer to them as follows:

Option 1. The Status Quo (the current Superfund program)
Option 2. Expanded Mixed Funding for Orphan Shares
Option 3. Liability Release for All Closed Co-disposal Sites
Option 4. Liability Release for All Pre-1981 Sites
Option 5. Liability Release for Current NPL Sites

This is also an appropriate place to discuss briefly the criteria we employ to evaluate each of the five options. There are dozens of standards against which these options could be evaluated. Nevertheless, the dissatisfaction with the present statute revolves primarily around the slow pace of cleanups, the transactions costs it generates, and its perceived unfairness, and so these concerns are logical criteria to use in evaluating possible alternative liability standards. Additionally, the current liability provisions are thought by some to induce cleanups at sites not on the NPL and also to provide incentives for careful handling of hazardous substances that go beyond those provided by regulations under the Resource Conservation and Recovery Act (RCRA)—which pertains to a smaller universe of hazardous substances than Superfund does—and by other statutes. To reflect these two concerns, we also include voluntary cleanups at non-NPL sites and due care in future waste management practices as evaluative criteria. The final criterion is concerned with the financial implications of shifting cleanup responsibility from PRPs to the Trust Fund.

With regard to the last criterion, we do not attempt to trace out the full economic consequences associated with each alternative or the status quo. Rather, we estimate and discuss the increased revenues (assumed to come from an increase in the corporate environmental tax) required by the change in liability for each alternative. It is well beyond the scope of this report to make new and detailed estimates of the effects of these changes on individual industries or on the economy as a whole. Nor could we address here the possible effects of Superfund liability on the access of small enterprises to capital markets, or the possible chilling effects of Superfund liability on new start-ups in manufacturing. We hope to address these important subjects in future work.

To summarize, then, the six evaluative criteria, discussed in the last section of this chapter, are as follows:

1. Speed of cleanup
2. Transactions costs
3. Voluntary cleanups at non-NPL sites
4. Due care in future waste management practices
5. Fairness
6. Financial implications

We call attention to the fact that the degree of protection afforded by cleanups is not included in this list. It is omitted because we assume throughout our analysis that the standards for cleanup stay the same under each of the five options. That is, we assume for all of our options that there is no change in CERCLA Section 121, which provides guidance for remedy selection.

Of course, different financing options could create pressures to modify the degree—and cost—of cleanups. For example, it is possible that a public works program, under which all cleanups would be paid for by the Trust Fund, could lead to unnecessarily expensive cleanups because of the ready availability of cleanup dollars. Conversely, an undercapitalized Trust Fund could lead to shoddy cleanups. We recognize that a change in Superfund financing could have these important effects, but we do not address them because we want to focus here solely on financing issues—who is liable and how much in taxes is needed for the Trust Fund under different liability schemes.

We turn now to a fuller description of the liability options we consider.

Options

The five options are described here only briefly since their strengths and weaknesses are vetted fully in chapter 3. The options are not fleshed out as bona fide legislative proposals with specific statutory language in this report. We have purposely left many details vague—for example, the precise definition of "orphan share" and other issues with which Congress would have to wrestle. Our purpose here is to suggest the broad outlines of a few liability options and to trace out the likely effects of these changes. Each could and should be scrutinized, and each would need many details spelled out if it were to be enacted into law.

Option 1. The Status Quo

Under the current Superfund law, a wide variety of parties may be held liable for the costs of cleaning up sites contaminated by the actual or potential release of hazardous substances.[1] These parties include past and present owners and/or operators of the sites, the individuals or companies that arranged for the transportation of the hazardous substances to the sites, and those who arranged for disposal or treatment of the substances. Industries, municipalities, lending institutions like banks and savings-and-loans, small businesses, and even schools may fall into the category of liable parties as defined by the statute.

Superfund liability is strict, joint and several, and retroactive. *Strict liability* means liability without fault; thus PRPs may be held liable even if they can demonstrate that they had been managing hazardous substances in a prudent, non-negligent, and legal manner. *Joint and several liability* means that, generally, any one PRP could be held responsible for the entire cost of a site cleanup even if that party had contributed but a portion of the substances at the site (although that party could in turn sue to recover costs from other responsible parties, and this is called a "contribution action"). Finally, Superfund liability is *retroactive* in that it applies to contamination resulting from activities that took place prior to passage of the law in 1980.

Congress gave EPA two tools for getting sites cleaned up: the first is its powerful enforcement authority to issue orders or bring lawsuits (that also result in bringing PRPs to the table to negotiate with EPA); the second is the Trust Fund that EPA may use to finance cleanups itself (either at "orphan sites" where no PRPs can be found, or at other sites in order to initiate prompt cleanups for which costs would eventually be recouped from PRPs).[2]

The Trust Fund is stocked with revenues from a variety of sources: these include a tax on domestically produced and imported oil (about $570 million per year); a similar tax on feedstock chemicals (about $245 million per year); the aforementioned corporate environmental tax (about $460 million per year); general revenues (intended to be $250

---

[1] There are a number of in-depth descriptions of the Superfund program. See, for example, Jan Paul Acton, *Understanding Superfund: A Progress Report* (Santa Monica, Calif., RAND Corporation, The Institute for Civil Justice, 1989); Clean Sites, *Making Superfund Work* (Alexandria, Va., January 1989); U.S. Congress, Office of Technology Assessment, *Superfund Strategy* (Washington, D.C., OTA, April 1985).

[2] Throughout this report all government action, including enforcement, is referred to as stemming from EPA, although in many cases the U.S. Department of Justice and state agencies are also involved.

million per year, but unappropriated for some years); and a variety of other sources, including interest on the Trust Fund balance and cost recoveries from PRPs. In Fiscal Year 1990, total Superfund revenues were $1.53 billion.[3]

Option 2. Expanded Mixed Funding for Orphan Shares

EPA has the option of paying for (or carrying out) a portion of site cleanup activities itself by holding settling parties liable only for their portion of the costs at the site (this is known as "mixed funding"). EPA may then attempt to recover its costs from PRPs not party to the settlement agreement (nonsettlors).

Option 2 would have EPA employ mixed funding much more aggressively to cover *all* orphan shares as well as the shares of any nonsettlors (or recalcitrants) at any NPL site. Although the liability standards in the law would remain unchanged, as a matter of policy financially viable PRPs would not be pursued "jointly" to force them to pay for all cleanup costs at sites. This option would require the development of a workable definition of "orphan share," a framework for allocating cost shares (for example, that the allocation be based on volume or toxicity or risk, and so on), and active cost recovery for the shares of recalcitrants. Under this option, EPA would continue to identify PRPs on a site-by-site basis.

Option 3. Liability Release for All Closed Co-disposal Sites

Municipal solid waste landfills, at which many generators—public and private—have disposed of wastes (this is referred to as co-disposal), are among the most contentious on the NPL.[4] From a practical standpoint, difficulties arise because of both the sheer volume of wastes at many of these sites and the difficulties of linking particular PRPs with specific wastes. Even where the volume of hazardous substances contributed by private firms and municipalities is small in relation to the total volume of wastes at a site, cleanups can be extraordinarily expensive if the whole site requires remediation. These sites also are difficult from a legal standpoint because they can involve thousands of PRPs, such as the 4,000 identified at the Operating Industries, Inc., site in Los Angeles. In cases where EPA does not name all the PRPs at the site, these parties may be named in third-party contribution actions (intra-PRP litigation). Thus, these are the sites at which cost allocation disputes—and hence initial transactions costs (primarily legal fees)—and intra-PRP litigation are likely to be the greatest.

With an eye primarily to addressing the latter of these two problems (high transactions costs), under Option 3 liability would be waived for *all* PRPs, public and private, at landfills at which industrial and municipal solid wastes were co-disposed and that have ceased operation (that is, closed) before the date on which Superfund is amended. This liability waiver would apply to sites listed on the NPL in the future so long as they meet the

---

[3] Based on information provided for FY 1990 by the Office of the Comptroller, U.S. Environmental Protection Agency, Washington, D.C., April 1992.

[4] We exclude from this definition a landfill owned and operated by a municipality that *only* manages household wastes of people living in that municipality.

same criteria. This release from liability would *not* apply to co-disposal sites that close subsequent to the passage of an amended act.

Option 4. Liability Release for All Pre-1981 Sites

One of the most common complaints about Superfund concerns its retroactive imposition of strict liability. PRPs repeatedly point to sites for which they are now liable but at which no hazardous substances have been deposited for many decades and where, they say, all waste management was in compliance with the rules in place at the time.[5] Although the courts have left little doubt about the constitutionality of retroactive liability, its application has embittered many PRPs.

Under Option 4, liability would be waived for all PRPs at NPL sites at which the disposal operation closed prior to the passage of Superfund in December 1980 (for convenience, we select a cut-off date of December 31, 1980, and refer to these as "pre-1981 sites").[6] This liability waiver would also apply to sites listed on the NPL in the future so long as they met the same criterion. Under Option 4, PRPs would continue to retain Superfund liability at facilities at which the disposal operation remained open after December 31, 1980—even if disposal at these facilities was in accord with current regulations and even if some hazardous substances at these sites were deposited before January 1, 1981.

Some in the Superfund policymaking community have suggested earlier dates as an appropriate cut-off point—when the Resource Conservation and Recovery Act (RCRA) was passed, for example. Others have argued for a later cutoff date—when major RCRA regulations became final in 1984, for instance. We selected the date of Superfund's enactment for Option 4 because no PRP could have known that it would be held liable under Superfund before 1981 and because after that date, it was the PRP's responsibility to know. Clearly, if Congress considers amending Superfund to include a release from liability based on a cut-off date, much debate will center on just when that date should be.

Option 5. Liability Release for Current NPL Sites

Although it would not prove convincing to all, a case can be made that if the primary goal of Superfund is to "get sites cleaned up," EPA should stop wasting time finding responsible parties and making them pay for cleanup. Instead, the agency itself should clean up all those sites that warrant federal action. Accordingly, we include as a final option one in which all sites on the NPL at the time the Superfund law is amended would be cleaned up by EPA. That is, the current liability standards would only apply to those sites added to the NPL in the future. While the logic here might seem to suggest that all future NPL sites be cleaned up by the Trust Fund, we do not evaluate such an option because it would eliminate any incentive for prudent waste management in the future and could be a huge financial drain on those paying the taxes to support the Trust Fund.

---

[5] In some cases, however, the cause of contamination was waste disposal practices that were *not* in accord with the practices of the time (see appendix A in this report).

[6] Other ways of defining what it might mean to be a "pre-1981" site are discussed in chapter 3.

Criteria

We chose six benchmarks against which to evaluate the options discussed above. These criteria are described next.

Speed of Cleanup

One of the most universal criticisms of the current Superfund program is that cleanups are slow to begin and slower still to be completed. There is less agreement about why this is so, about how to accelerate cleanups without sacrificing quality, and about whether sites would be cleaned up much more quickly absent the negotiation and litigation that now take place.

At least three major activities affect the length of time between the date a site is put on the NPL and the date when the final remedial action is completed. The first activity is the site study phase—the time needed to assess the contamination at the site, develop possible remedial alternatives, select an appropriate remedy, obtain public input on that approach, and design the final remedy. The second is the actual cleanup process—the time it takes to implement the selected remedial plan, whether that plan involves construction of an on-site incinerator, a cap on a landfill, installation of a leachate collection system, excavation of soils, or some combination of these (and possibly other) actions. The third activity is the settlement and enforcement process—the time required to reach agreements with PRPs to conduct site studies and cleanup and/or to bring enforcement actions against those who do not enter into agreements with the government to perform this work. This third step is generally undertaken concurrently with site study and cleanup activities.

The speed with which these efforts take place can be affected by many factors, including the technical complexity of the site, the concerns of local citizens, disagreements between a state and EPA, and the number of responsible parties involved. In addition, availability of adequate funds—whether from the Trust Fund or from PRPs—affects how many sites EPA can move through the cleanup process at one time.

It is important to note that EPA recently announced a new strategy that is intended to hasten cleanups and to present a clearer picture to the public of which sites will take more than five years to remediate.[7] The broad outline of EPA's new cleanup paradigm—which blurs the current distinction between the "short-term" *removal* program and the long-term *remedial* program—was made public in March 1992. It is too soon to tell what effect this new strategy will have on the cost or speed of site cleanups if it is implemented. Because the new cleanup strategy has only recently been unveiled, our analysis is based on information for the existing program. If the new approach is fully and successfully implemented, however, it could reduce the cost estimates for site cleanup employed in this report as well as the amount of time needed for cleanups.

It is important in evaluating any alternative to the current Superfund to distinguish between those delays that are inherent in each liability and taxing scheme and those resulting from implementation policy and procedures. In this report, we do not suggest

---

[7] U.S. Environmental Protection Agency, Office of Emergency and Remedial Response, "Superfund Accelerated Cleanup Model" (Washington, D.C., EPA, March 1992).

ways to improve the implementation of the site study or cleanup processes. That has been done in many previous reports.[8] Although we also avoid speculating about ways to improve the overall settlement and enforcement process, Option 2 (Expanded Mixed Funding for Orphan Shares) would represent a fundamental change in EPA's settlement policy.

To assess how each option will affect speed of cleanup, we discuss the impact of each on the speed with which responsible parties are likely to reach a settlement with EPA[9] and agree to take the lead for cleanup activities. We speculate only about the total time between NPL listing and completion of the last remedial action. We do not attempt, as others have done, to look at the length of individual steps in the cleanup process, such as the average duration of the remedial investigation/feasibility study (RI/FS).[10]

Some of the options considered in this report would increase substantially the number of sites at which EPA would be directly responsible for site studies and cleanup activities. Without a more comprehensive analysis of EPA's staffing and internal site management structure, it is difficult to know just how many sites the agency could effectively manage at one time, although it could perhaps manage the same number of sites now being actively worked on by EPA and PRPs combined because it now conducts oversight of all PRP-lead activities anyway. In FY 1991, for example, 13 percent of EPA's total Superfund resources were slated for site enforcement activities and could have been reassigned to site management functions if the lead for cleanup at many sites had been shifted to EPA.[11]

EPA probably would need to develop new internal organizational systems and also reassign and train staff if it was to clean up 600 to 1,200 sites by itself, however. This would almost inevitably create some delay, if only during the transition from the current to a new liability scheme. Another important issue is whether a public works program, such as Option 5, having no PRP involvement, might take longer because of the lack of external attention to the program. We do not address these issues in our analysis, but note here that they are critical in any consideration of alternative liability approaches.

Transactions Costs

A variety of transactions costs arise under Superfund. The most prominent are those stemming from negotiations of settlements and preparations for and the actual conduct of litigation. At a single site, at least five separate tiers of litigation/settlement negotiations can

---

[8] Acton, *Understanding Superfund*; Clean Sites, *Making Superfund Work*; Office of Technology Assessment, *Superfund Strategy*; U.S. Environmental Protection Agency, *A Management Review of the Superfund Program* (Washington, D.C., EPA, June 1985).

[9] Under Superfund, a state can be the government agency in charge of supervising cleanup or enforcement actions in lieu of EPA, but for ease of reading here, we refer simply to EPA.

[10] See testimony of Jan Paul Acton, Congressional Budget Office, at hearings on Administration of the Federal Superfund Program, before the Subcommittee on Investigations and Oversight of the House Committee on Public Works and Transportation, 102d Cong., 1st Sess. (1991).

[11] U.S. Environmental Protection Agency, Office of Emergency and Remedial Response, *Progress Toward Implementing Superfund, Fiscal Year 1990: Report to Congress* (Washington, D.C., EPA, February 1992), p. 35.

occur: (1) between EPA and PRPs, (2) among settling parties, (3) between settling parties and nonsettlors, (4) between PRPs and their insurers, and (5) between insurers and re-insurers.[12] At each site there may be many PRPs, all of whom may hire lawyers and technical experts to represent their interests. In addition to legal costs, PRPs often conduct their own risk assessments or site evaluations to provide an independent assessment of EPA's work.

Some believe that EPA could reduce some private-sector transactions costs—those resulting from EPA-PRP negotiations and also intra-PRP litigation—by employing more aggressively the settlement provisions outlined in CERCLA Section 122: using mixed funding, *de minimis* settlements (that is, settlements with small parties), and nonbinding allocations of responsibility (NBARs); naming *all* PRPs in the initial settlement negotiations; and giving settling parties protection from contribution actions, where appropriate.

Under Options 2 through 5, some PRPs would be released from liability. In these cases, three new considerations arise, one of which implies possible new transactions costs. Although these costs are likely to be small, they are worth noting. The first consideration involves litigation about whether a site qualifies for liability release. Any line drawn to separate those who pay from those who do not is likely to foster litigation. For instance, if it was decided that all sites that closed before a certain date were to be cleaned up with monies from the Trust Fund, there would likely ensue litigation over the closing dates of at least some sites. This litigation and the costs it would engender could be kept to a minimum by specifying explicitly the conditions for liability release and fixing clearly the burden of proof.

The second consideration has to do with one of the effects of cleanups financed by the Trust Fund. Specifically, they could enhance the value of the affected sites. If so, when these sites were sold, the owner would reap a kind of unearned or "windfall" profit. Thus, some mechanism—a lien provision, perhaps—would have to be established to assure that the government would be reimbursed for past cleanup costs if such sales took place. Such a provision, which would require legislative action, would give rise to some new administrative costs, but those costs would likely be negligible.

The third consideration, to which "new" transactions costs might be attached, results from the transition from one liability approach to another. If PRPs were released from liability for sites on the NPL, those who already settled with EPA for past cleanup or site study costs under the current liability regime might demand reimbursement for expenses incurred at those types of sites where liability is no longer being imposed. In cases where EPA was reimbursed for past costs, paying PRPs back might be administratively feasible, although it would be necessary to establish procedures for reimbursement where parties have reallocated these costs through third-party suits. However, where PRPs have taken the lead for conducting site studies and cleanups, compensating them would be more difficult because EPA has not required PRPs to monitor and report the cost of cleanups. Because relatively few sites have reached the actual cleanup stage, Congress might well decide that not reimbursing PRPs for past cleanup costs is a small price to pay for assuring a more orderly change to a new system. This issue should be directly addressed by legislation to avoid confusion if substantive changes are made in the liability standards.

---

[12] The magnitude of insurer–re-insurer transactions costs will depend in large part on the result of PRP-insurer coverage litigation.

We evaluate both the current program and the alternative approaches we have selected according to their effect on transactions costs, including (occasionally) the possible new types discussed immediately above. Our analysis is mostly qualitative in nature, and we have conducted no original research on these costs. We do, however, make a ballpark estimate of likely reductions in transactions costs for each option; we also rely on a recently issued report by RAND on transactions costs in our analysis of the current program.[13]

Voluntary Cleanups at Non-NPL Sites

To some, one of the important benefits of the Superfund program is that, because of concerns about future liability, PRPs may choose to clean up sites not currently on the NPL. These "voluntary" cleanups could take the form of remediation by the owner or operator of sites prior to or during real estate transactions; cleaning up contaminated areas even where no sale is contemplated; and cooperation by PRPs (who may or may not be the owner/operator) with other PRPs and the relevant state agency to clean up non-NPL sites under state Superfund laws.

Unfortunately, little is known about the effects of the current law on such non-NPL cleanups. Thus, we can only speculate—and we chose this word carefully—about whether the options analyzed here would provide incentives for PRPs to take such actions. To the extent that they exist, these incentive effects will vary depending on whether a site is single-party (that is, it was used solely for on-site disposal by the owner or operator of the site) or multiparty (such as a municipal landfill, where many generators are involved). Some sites fall between these categories; they involve more than one party because there was a change in ownership during the life of the site, but they do not involve many parties. While these are not strictly single-party sites, we lump them in this category since responsibility for contamination is confined to a reasonably well-defined set of PRPs. In some cases, the options we consider may actually create incentives for PRPs wanting sites on the NPL to obtain EPA-financed cleanups.

Due Care in Future Waste Management Practices

Liability rules have at least two purposes; not only do they help allocate the costs associated with previous actions, but they also can provide incentives to avoid behavior that would give rise to future risks—in this case, those arising from improper management of hazardous substances. RCRA and other federal laws, state laws, and regulations; the threat of toxic tort suits; and the natural desire to be viewed as a responsible member of society all provide incentives for the careful management of hazardous substances. However, because Superfund liability attaches to such a broad category of hazardous substances, its proponents argue that it provides an added (and perhaps even more powerful) incentive for those dealing with hazardous substances to reduce and recycle as much as possible and to manage properly what is left over.

---

[13] Jan Paul Acton, Lloyd S. Dixon, and others, *Superfund and Transaction Costs: The Experiences of Insurers and Very Large Industrial Firms* (Santa Monica, Calif., RAND, The Institute for Civil Justice, 1992).

The most contentious issue relating to these "incentive effects" is the role of retroactive liability. Virtually everyone would agree that prospective liability affects a PRP's behavior. It is much less clear whether the establishment of liability for *past* actions could or would materially affect the future waste-handling practices of public or private parties. We therefore discuss the likelihood that the current program induces more careful management practices; we also speculate about the likely impacts of alternative liability approaches on incentives for due care.

Fairness

One of the most controversial of the evaluative criteria we consider, fairness is also the most difficult to define.[14] For instance, in the environmental arena fairness is often equated with (or said to be exemplified by) the so-called "polluter pays" principle, which holds that those who create environmental hazards are responsible for ameliorating them and/or paying for the damages they engender. Indeed, this principle is one on which the environmental laws and regulations of all the countries of the Organisation for Economic Co-operation and Development (OECD) are supposed to be based.

However, fairness can have other connotations, as well. For example, some would view an environmental liability standard or other policy as fair if it accorded equal treatment to parties which had handled hazardous substances in similar ways. This would be analogous to the doctrine of horizontal equity in tax policy—that is, treating equals equally. Finally, fairness could be construed to apply to more than the identities of those bearing costs; in other words, it could also pertain to the consistency of any set of rules or procedures. In this spirit, fairness could reasonably be interpreted to mean that changes in liability standards would only be made prospectively; that is, changes in the "rules of the game" would not be made retroactively.

Fairness is still more complicated in the case of Superfund because one must consider not only who is liable for cleanup, but also who pays the taxes that contribute to the Trust Fund. Eliminating liability for some PRPs may seem unfair unless those PRPs contribute significantly through Superfund taxes to the Trust Fund, from which the added revenues for cleanup will come. In this report, we concentrate on the fairness of the various liability standards we consider, but we also on occasion make reference to the taxing mechanism.

Financial Implications

It is clearly of interest to know the financial implications for the Trust Fund of possible changes in the liability standards. For each of the alternative approaches considered, we estimate how much additional revenue would have to be added to the Trust Fund to implement the option, assuming that the current pace of the program is maintained. For example, in estimating the financial implications of waiving PRP liability for pre-1981 sites,

---

[14] This section is based loosely on an essay by Paul R. Portney in the *EPA Journal* vol. 17, no. 3 (July/August 1991), pp. 37-38.

we estimate how many such sites are on the current NPL and what additional monies would be needed if the Trust Fund was to be used to pay for their cleanup.

Given the lack of hard data on actual site cleanup costs, we make a set of assumptions regarding the basic components of the Superfund program that we use to estimate the annual and cumulative funding requirements of the current program and the four alternative approaches we consider. **Changing any of these assumptions would have a major impact on our estimates of both the annual and cumulative costs of the program.**

The assumptions underlying our calculations are as follows:

1. We assume that the average cost of a remedial action (RA) will be $40 million per site. Since EPA estimates RAs to cost $27 million per site, our estimate of the cost of the current program is higher than EPA's.

2. We assume that the total number of sites on the current NPL that are relevant to our analysis is 1,115. This was the number of sites (other than federal facilities) on the NPL when we received site-specific information from EPA in July 1991. We do not calculate the costs associated with the 100 new sites EPA expects to add to the NPL each year because of the difficulty of identifying how many of these sites would be affected by each of the liability options we consider. Note, however, that at 100 new NPL sites a year, 1,000 new sites would be added in the next 10 years, nearly doubling the Superfund program; this would imply additional remedial action costs of $40 billion.

3. We accept EPA's estimate of approximately $1.0 billion as the annual cost of implementing all aspects of the Superfund program other than EPA's costs for implementing remedial actions.[15] Although in theory many of these costs can be recovered by EPA from PRPs, we do not address potential revenues from cost recovery. In addition, we do not estimate any change in Trust Fund revenues for the alternative approaches resulting from decreased cost recovery, because revenues from cost recovery to date have been insubstantial.[16]

4. Although the relative shares of the different components of EPA's expenditures for functions other than RAs (that is, for enforcement, site oversight, and so on) are likely to change over the years and for different policy options, we do not address such changes. To estimate such implications would require a sophisticated analysis of EPA's budget. In addition, over the long run the dominant factor in total Superfund cost is the cost of implementing site remedies.

5. Although it is optimistic, we assume that it takes 10 years from the time a site is proposed for inclusion on the NPL until the final remedy is fully implemented. Furthermore, for the sake of simplicity, we assume that remedial action costs are spread evenly over the 10 years even though this is not the pattern one observes in reality. We do so because we doubt that Congress would want to address appropriations levels that vary from year to year, considering the paucity of data on the magnitude and timing of site cleanup expenses.

---

[15] This figure is based on FY 1990 Trust Fund revenues of $1,533.9 million minus an estimated $500 million for RAs, per conversations with EPA staff in the Office of Emergency and Remedial Response and the Comptroller's Office.

[16] According to EPA's *Superfund FY 1991 Annual Management Reports* (Washington, D.C., EPA), as of September 30, 1991, cumulative cost recovery collections were $352.8 million.

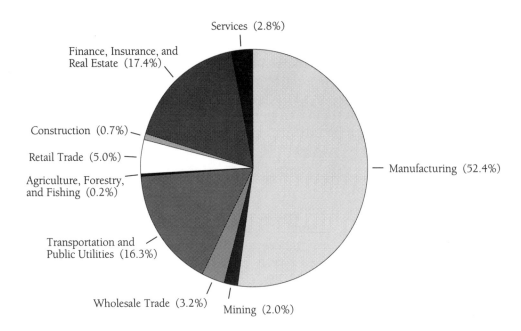

Figure 1. Share of corporate environmental tax paid by major industrial sectors in 1988.

Source: U.S. Department of the Treasury, Internal Revenue Service, *Statistics of Income—1988: Corporation Income Tax Returns* (Washington, D.C., U.S. Government Printing Office, 1991).

Using all these assumptions, we calculate the additional Trust Fund revenues for remedial actions (only) that would be needed to implement each of our four alternative approaches. We also calculate the "implied" PRP expenditures for RAs. This calculation is made by deducting planned EPA expenditures for remedial actions from our estimate of the total cost of cleanup for all sites.

After estimating the additional revenues that each of the four alternative approaches implies for the Trust Fund, we calculate how much the corporate environmental tax would need to be increased to raise this sum. We also identify how such an increase in the corporate environmental tax would affect the relative share of annual total Trust Fund revenues from key industry groups such as chemicals and allied products, petroleum and coal products, and insurance, to name a few (see appendix D).

In FY 1990, the corporate environmental tax provided just under one-third of all Trust Fund monies. This tax is levied at a rate of 0.12 percent of each corporation's alternative minimum taxable income (AMTI) in excess of $2 million, regardless of whether that company is actually subject to the alternative minimum tax.[17] The corporate tax is much more broadly based than the petroleum and chemical feedstock taxes, with the manufacturing sector paying just over 50 percent of the tax in 1988, the latest year for

[17] The alternative minimum tax (AMT) was created in 1986 to ensure that taxpayers with substantial incomes could not avoid paying taxes through the (legal) use of deductions, tax credits, and other exclusions permitted under the law.

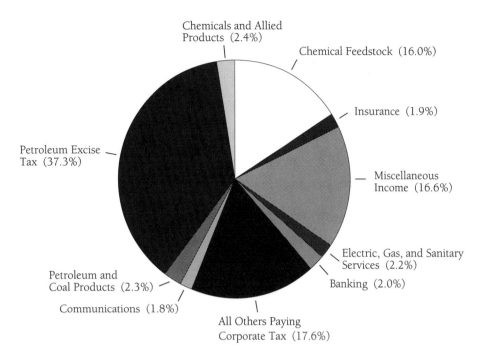

Figure 2. Share of total Superfund revenues for 1990.

Note: Information on total revenues is for 1990. The breakdown of the corporate environmental tax into specific industries, such as banking, is based on 1988 figures, the last year for which they are available.

Source: Appendixes B and D in this report.

which this information is available.[18] (See also appendix B for detailed information on who paid the corporate environmental tax in 1988.) Figure 1 shows how much of the corporate environmental tax was paid by major industry sectors in 1988.

Three major industry groups—electric, gas, and sanitary services; chemicals and allied products; and petroleum and coal products—paid 23 percent of the total corporate environmental taxes for 1988. The next three highest-contributing industry groups were insurance, banking, and communications—paying 19 percent of the tax. Thus, six industry categories paid just under 42 percent of the total corporate environmental tax.

To provide a better picture of the overall financing of the Trust Fund, figure 2 presents information on the chemical and petroleum feedstock taxes, the top six contributors to the corporate environmental tax, and miscellaneous revenues.

---

[18] All information on the corporate environmental tax in this report is based on U.S. Department of the Treasury, Internal Revenue Service, *Statistics of Income—1988: Corporation Income Tax Returns* (Washington, D.C., U.S. Government Printing Office, 1991).

3

Analysis of Policy Options

Each of the five policy options that we address—namely, the liability standards in the current Superfund program (the "status quo") and the four alternative approaches identified in chapter 2—has strengths and weaknesses. Depending on the weights assigned to the evaluative criteria—reducing transactions costs, providing incentives for responsible waste management, and so on—different liability and taxing schemes appear more or less attractive. In the following sections, we evaluate each of the options against each of the criteria identified in chapter 2.

Option 1: The Status Quo

Because Option 1 forms the baseline against which the alternative approaches are evaluated, the analysis of the current program is somewhat more involved than that of Options 2 through 5. This is especially true of the discussion of financial implications, where we describe in some detail the economic consequences of Superfund as it has been implemented.

Speed of Cleanup.   Virtually everyone would agree that cleanups take a long time under the current Superfund program. Recent testimony by officials of the Congressional Budget Office suggests that for those sites that ultimately end up on the National Priorities List, the total elapsed time between a site's first being brought to EPA's attention for evaluation and the completion of the remedy may average as much as 15 years.[1] According to information from EPA's central database (CERCLIS—the Comprehensive Environmental Response, Compensation, and Liability Information System), it takes nearly 11 years, on average, from the time a site is first proposed for inclusion on the NPL to the time EPA expects the last currently planned remedial action to be completed.[2]

---

[1] See testimony of Jan Paul Acton, Congressional Budget Office, at hearings on Administration of the Federal Superfund Program, before the Subcommittee on Investigations and Oversight of the House Committee on Public Works and Transportation, 102d Cong., 1st Sess. (1991).

[2] This information is based on the authors' analysis of CERCLIS information provided by EPA in July 1991.

Even the latter figure is almost certainly an underestimate, because the CERCLIS database only includes site activities scheduled through 1999. EPA's pledge in October of 1991 that 650 of the 1,250 sites now on the NPL will be cleaned up by the year 2000 means that at 600 sites all remedial activities will not be completed by then.[3] Moreover, many sites will require groundwater treatment for years even after the remedy has been implemented. If EPA completes remedial actions at the projected rate of 60 per year, the last site on the *current* NPL would be cleaned up in 2010—18 years from now.

Although these data on time-to-completion are disturbing, they do not help pinpoint the role the present liability standards play in these delays, as compared with other factors. Certainly the need to identify potentially responsible parties and to get them to agree to conduct and finance both site studies and cleanups results in some delay that would not happen if EPA had adequate monies in the Trust Fund to pay for all cleanups. There are, however, other factors apart from the liability system that can delay cleanup. They include the following: the site study and remedy selection process that EPA has established, which requires a thorough investigation of site contamination and a comprehensive analysis of possible remedial actions, and which can be time-consuming even at relatively simple sites; the need to coordinate with the appropriate state agencies; and the need to address the concerns of local citizens regarding the level of cleanup to be achieved and how cleanup will be implemented. In addition, implementing the remedy itself can take years.

Thus, both the liability scheme and the nature of the site study and remedy selection process contribute to slow cleanups. While a number of studies have looked at how these factors affect the speed of cleanup at individual sites, no meaningful generalization about the largest overall contributor to delay is possible.[4]

In addition, some of the causes of delay in cleanups associated with the current liability standards in Superfund are a consequence of policy choices made by EPA rather than of the standards per se. For example, the current law does not preclude EPA from cleaning up sites itself and then recovering money from private parties after the fact. This strategy would eliminate what many believe to be a major source of delay—the effort required to obtain funds up front from PRPs. In fact, EPA adopted just such a "Fund first" policy during the early years of the Superfund program, but was widely criticized at the time for wasting Superfund money and letting PRPs off the hook. In response, the agency adopted its current "enforcement first" strategy in an effort both to increase the number and dollar value of PRP settlements and to leverage Trust Fund monies more effectively. It is still too soon to know what effect this strategy has had on speed of cleanups. Striking the right balance between using the Trust Fund to speed cleanups and "enforcing first" to leverage Trust Fund dollars is no easy task.

Similarly, many argue that EPA could streamline the site study and evaluation process and thereby greatly accelerate cleanups. As noted in chapter 2, EPA proposed just such a strategy in March 1992, also announcing plans to develop "model remedies" for certain

[3] Bureau of National Affairs, "New Superfund Head, Trouble Shooter Staff Named to Review Contracts, Speed Cleanup," *Environment Reporter*, Oct. 4, 1991 (Washington, D.C., BNA), p. 1405.

[4] See, for example, *Superfund Site Studies* authored by National Strategies, Inc., for American International Group, Crum and Forster, and Fireman's Fund Insurance Companies; and Thomas W. Church, Robert T. Nakamura, and Phillip J. Cooper, *What Works? Alternative Strategies for Superfund Cleanups* (Washington, D.C., Clean Sites, September 1991).

kinds of sites and contamination problems (generic approaches to site cleanup that could be applied quickly).

Finally, it is possible that eliminating the need to bring PRPs to the negotiating table might not, in the long run, speed cleanup. Many in both the business and environmental communities have questioned whether a full public works program for all sites on the NPL would be any quicker or more efficient than today's system. The most often-voiced concern is that EPA would select the most costly ("Cadillac") cleanups, resulting in much higher taxes to finance these gold-plated remedies. Others argue that the reason cleanups are so slow may be the lack of consensus about cleanup standards alluded to earlier, an issue not addressed here.

In summary, cleanup is slow under the present liability regime in Superfund. It is a matter of great contention whether this is due primarily to the existing liability scheme or to the way in which the law is implemented. There are many opportunities to speed cleanup: it could be accelerated by changing the liability standards (*assuming* adequate Trust Fund resources and agency staff), by streamlining the site study and remedy selection process, or by implementing a different settlement and enforcement policy. Deciding which of these options is "best" depends on one's views of the goals of Superfund.

Transactions Costs. Several kinds of transactions costs arise under the current law. The most frequently cited are those resulting from legal actions. These include settlement negotiations between the government and the PRPs; litigation by the government against PRPs; cost allocation negotiations among PRPs; litigation among PRPs; litigation between PRPs and their insurers; and, eventually, litigation between insurers and their re-insurers. EPA has recently stepped up its efforts to get small ("*de minimis*") parties out of the process early in order to reduce transactions costs and to help speed up settlements. The agency is also evaluating the promise of other settlement tools. These efforts could reduce some PRP-related transactions costs, but would do little—perhaps nothing—to reduce insurance-related transactions costs.

Many also argue that the level of detail in site studies is driven more by legal than by engineering concerns, adding both time and expense to the site study process. In addition to these types of litigation-related expenses, when EPA has the lead for cleanup activities, many PRPs conduct their own site studies to gather independent information on both the contamination at the site and on the feasibility and cost of alternative remedial actions. This results in what is often an unnecessary and expensive duplication of effort.

Until recently, there was little comprehensive and reliable information on the magnitude of transactions costs. However, a new report by RAND documents transactions costs at Superfund sites for five Fortune 100 industrial firms and four national insurance carriers.[5] This report sheds much-needed light on the sources and magnitude of these costs.[6]

According to RAND researchers, the percentage of Superfund outlays accounted for by transactions costs varies greatly between PRPs and insurers. They found that transac-

[5] Jan Paul Acton, Lloyd S. Dixon, and others, *Superfund and Transaction Costs: The Experiences of Insurers and Very Large Industrial Firms* (Santa Monica, Calif., RAND, The Institute for Civil Justice, 1992).

[6] RAND has also initiated a study, funded by U.S. EPA's Office of Policy Analysis, examining transactions costs for small and midsized PRPs.

tions costs for the five large PRPs in their survey averaged 21 percent of total Superfund outlays for all the sites at which the PRPs had involvement (including those in the early stages of the process). In 1989, the average total annual outlay for all Superfund-related costs for these five PRPs reached $6.1 million, almost three times higher than it had been three years earlier. Even so, the authors found this to be "surprisingly low" considering that the average number of Superfund sites per company was 144.[7]

RAND's researchers also found that the transactions cost share (that is, the percentage of total Superfund outlays going to transactions costs) varied greatly depending on the number of parties involved at the site and also on the site's stage in the cleanup process. According to RAND, transactions cost shares were very low at single-party sites (7 percent) and were 34 percentage points higher at multiparty sites. Transactions cost shares were also lower at sites that were further along in the cleanup process (that is, those where remedial actions were well along or had been completed).[8]

The four insurers in RAND's survey spent a combined $70 million on Superfund claims and related expenditures in 1989, double their 1986 outlays. Most of this money—almost 90 percent—went to defending their policyholders in legal actions with EPA, states, or other PRPs, and for disputing (with these same policyholders) whether their policies covered the cleanup costs. This percentage was lower (69 percent) for closed claims.[9]

The RAND report provides much-needed data on transactions costs. Nevertheless, it raises as many questions as it answers because of the relatively small number of firms surveyed. In addition, because the information was gathered by company—not by site—it makes it difficult to use RAND's findings to estimate the total amount of money going to transactions costs as compared with that for site remediation.

Reducing or eliminating each of the different types of transactions costs—those between EPA and PRPs, those among PRPs, and those relating to insurance coverage— may require slightly different solutions. For example, transactions costs resulting from what PRPs believe are unfair allocations of responsibility (costs that would result in delay during PRP-EPA negotiations and litigation among PRPs) could be reduced if EPA would bring more PRPs into the settlement and enforcement process and would use more aggressively the settlement tools provided by Congress in CERCLA Section 122. Another provision already noted allows EPA to get more parties out of the negotiation process, by dispensing quickly with small ("*de minimis*") contributors to a site. This reduces the number of parties to be dealt with and eliminates potentially large transactions costs for *de minimis* parties. As suggested in chapter 2, CERCLA Section 122 also allows EPA to pay for a portion of a PRP-conducted cleanup itself—through so-called mixed funding. Finally, Section 122 allows EPA to develop a nonbinding allocation of responsibility (NBAR) to suggest how costs should be apportioned among the PRPs. In some circumstances EPA has the authority to provide a PRP—as part of a settlement agreement—with protection from third-party "contribution" suits. These contribution actions, when one PRP who has incurred response costs sues another PRP (or many others) for "contribution"—that is, for their share of the costs incurred—are becoming more common. Implementing any or all of these provisions

---

[7] Acton and others, *Superfund and Transaction Costs*, pp. 33 and 35. RAND defined Superfund sites as including both NPL and non-NPL sites.

[8] Ibid., pp. 51-54 and 57.

[9] Ibid., pp. 20, 24, 26, and 27.

would shift some transactions costs from the PRPs to EPA. For example, at most sites the PRPs negotiate among themselves to develop a cost allocation. If EPA were to employ NBARs more often, the agency would then bear some of these transactions costs itself.

Many argue that utilizing all these settlement tools would result in a fairer allocation of costs among responsible parties. This in turn would give private parties a greater incentive to agree to conduct cleanups (and perhaps to agree more quickly as well). Still, EPA has had these settlement tools at its disposal for some years now, but with the exception of the *de minimis* provisions they have been little used. According to an analysis prepared by attorneys at Beveridge & Diamond, these provisions have not been utilized more often primarily because of limited staff training and experience, limited financial resources, and the low priority accorded to such efforts at EPA.[10] As for insurance coverage litigation, little can be done to reduce these transactions costs except by releasing categories of PRPs from liability, at least until this issue reaches the highest court in each state and legal precedent is made clear.

To summarize, we offer no estimate of the total transactions costs now being incurred as a result of the current Superfund liability standards. Data compiled by RAND, however, suggest that total transactions costs are *probably* measured in hundreds of millions of dollars annually.[11] We hasten to add three important caveats, however. First, this conclusion is arrived at much more through speculation than by estimation. Second, not *all* transactions costs result from the liability standards per se; some are no doubt due to other aspects of the implementation of the law. Third, RAND's findings on transactions and cleanup costs pertain not only to sites on the NPL but to other sites as well. This raises the possibility that Superfund liability and the costs it entails act as a stimulus to clean up sites not yet on the NPL, an issue to which we now turn.

Voluntary Cleanups at Non-NPL Sites. The incentive to undertake voluntary cleanups at sites not on the NPL could take a number of forms: encouraging private parties to clean up sites they think might be placed on the NPL to avoid the red tape of the Superfund process; cleaning up sites to avoid future Superfund liability; cleaning up sites that are involved in property transactions because the prospective buyers are concerned about acquiring possible future Superfund liability; and developing or maintaining a record of corporate responsibility.

Unfortunately, there are few data documenting the number of non-NPL cleanups that are taking place, for whatever reason. Two recent studies suggest that such cleanups are occurring, although the data are very sketchy indeed. According to RAND, for example, six of the cleanups being conducted at the 73 sites in its survey where PRP expenditures were over $100,000 were self-initiated—three of these were at single-party sites and three at multiparty sites.[12] In addition, EPA and the Kennecott Corporation recently entered into an agreement under which the company will clean up hazardous waste sites at an operating facility without the agency's putting these sites on the NPL.[13]

---

[10] Beveridge & Diamond, P.C., "Superfund Transaction Costs: A Perspective on the Superfund Liability Scheme." Prepublication draft (Washington, D.C., 1991), pp. 35 and 37.

[11] Acton and others, *Superfund and Transaction Costs,* pp. 60-61.

[12] Ibid., pp. 47-48.

[13] Bureau of National Affairs, *Environment Reporter*, April 10, 1992 (Washington, D.C., BNA), p. 2696.

Even if it were possible to identify the precise number and type of cleanups at sites not on the NPL, it would be extremely difficult to document their cause definitively.[14] These cleanups might be due to concern about federal Superfund liability or state Superfund laws, to concerns over toxic tort liability, or simply to ethical concerns about environmental stewardship.

Some non-NPL cleanups may be the result of the corrective action provisions of Subtitle C of the Resource Conservation and Recovery Act (RCRA). It should be possible to pin these latter cleanups specifically to RCRA corrective action requirements, however, since the regulations pertain only to contamination at facilities that are regulated as hazardous waste treatment, storage, and disposal operations and that must obtain a permit under Subtitle C of RCRA.

The only aspect concerning actions at non-NPL sites about which there seems to be some agreement is that environmental audits—and sometimes cleanups—are increasing as a result of concern over Superfund liability arising in connection with property transactions. Beyond that, there is little consensus. Some we have talked to insist that any incentive created by the present liability standards pertains only to single-party sites, their logic being that only a site owner has an incentive to clean up that site truly voluntarily to avoid future liability under Superfund.[15] Others disagree, and believe that voluntary cleanups also are taking place at multiparty sites not yet on the NPL because PRPs want to avoid getting on the NPL and because the law allows PRPs to bring contribution actions to recover costs against other parties at these sites.

The debate about whether any incentive for voluntary cleanups affects only single-party (as opposed to multiparty) sites may be more semantic than real. There is little doubt that totally voluntary cleanups, absent any government involvement, are unlikely at a multiparty site such as a municipal landfill. However, there are multiparty sites where PRPs (usually the owner/operator) have worked cooperatively, usually with state or EPA involvement, to clean up sites not on the NPL.

We can offer no new information in this report on the nature and extent of non-NPL cleanups, but we find it difficult to conclude other than that Superfund liability does create an incentive to clean up sites not on the NPL and that retroactive liability in particular gives parties an incentive to clean up contamination that occurred before the passage of the Superfund law in 1980. If retroactive liability were eliminated, we believe it would diminish the incentive PRPs have to clean up "old" contamination at their facilities.

Due Care in Future Waste Management Practices. Despite a similar dearth of evidence with respect to this criterion, we also believe that strict, joint and several, and retroactive liability under Superfund provides an incentive to manage hazardous substances in an environmentally safe manner. Because of the "strict" nature of Superfund liability, the law gives individuals, companies, and government agencies an incentive to internalize (and reduce) likely future costs by managing hazardous substances carefully, even though the substances may not today be subject to a regulatory program such as RCRA. Joint and several liability gives generators an incentive to ensure that wastes are sent to well-

---

[14] Resources for the Future, *Summary Report: Superfund Research Symposium* (Washington, D.C., RFF, October 1990).

[15] There are issues about whether such "midnight cleanups" are actually legal, but we do not address them.

managed, technologically sophisticated treatment and disposal facilities—knowing that they could be held liable for cleanup of these facilities resulting from poor management of hazardous substances.

Difficult as it is to document non-NPL cleanups, it is even harder (if not impossible) to document how much "due care" and waste reduction are taking place in response to Superfund. Some qualitative surveys suggest that Superfund liability does encourage more careful management of hazardous wastes. For example, a report on reducing hazardous wastes in Wisconsin, based on in-depth interviews with 20 of Wisconsin's large and small generators of hazardous wastes, concluded that "the significant threat the liability provisions of SARA pose to a firm's long-term viability provides a strong incentive to reduce or eliminate hazardous waste."[16] As one survey respondent from a printing firm put it:

> We have waste ink of course which right now is not a hazardous waste. We're treating it as if it's going to be a hazardous waste, so we're taking it to a cement kiln. They run it through a secondary fuel and then they put the residue into cement as I understand it and then it's completely gone. There is no more. At one time we used to landfill it, years back. [Ibid., p. 18.]

Even if such changes in behavior could be measured accurately, attributing some fraction of them to Superfund liability would be problematic. For example, growing corporate commitment to waste reduction may—as we believe—be motivated in part by concern over Superfund liability. But such commitments could also be explained by the cost savings or other efficiencies that can sometimes result from them, by public pressure in response to the toxic emissions reporting required by Title III of SARA, by RCRA, by concern over toxic tort suits, and by other factors. Teasing out the "real" causes, much less apportioning their specific contribution to improved waste management practices, is extraordinarily difficult.

Virtually no one would dispute the assertion that prospective liability can and does have an effect on current and future waste handling and source reduction. The difficult question is whether *retroactive* liability also provides an incentive for due care, and, perhaps more importantly, what message would be sent if retroactive liability were eliminated. The argument against retroactive liability as a prospective incentive is based on the belief that holding parties responsible for past actions should not affect their future behavior. This argument has obvious appeal. But, in our view, the fact that liability has been imposed both retroactively and prospectively has forced those who use and handle hazardous substances today to pay attention to the fate of these substances.

The fact that parties are being held liable for actions they took in the past makes the strict liability of Superfund a reality. Perhaps most importantly, Superfund's powerful liability standards send a clear signal to PRPs that careless management of hazardous substances is not acceptable, even if they are in accord with all the rules and regulations at the time. It may not be the best way to get this message across, but it is an effective way. On the other hand, some PRPs have told us that Superfund does affect where they send their wastes for disposal, but—paradoxically—may not always lead them to select the "best" disposal facilities. For example, according to one PRP, Superfund liability creates an incentive for PRPs to continue sending wastes to sites where they have already "bought in"—that is, where they are already potentially on the hook for cleanup costs because of

---

[16] Department of Natural Resources, Bureau of Research, *Reducing Hazardous Waste in Wisconsin: A Summary of the Barriers and Incentives Wisconsin's Firms Encounter* (Wisconsin, March 1991), p. 17.

past disposal activities (although this could just increase their allocation at the site)—as opposed to "new sites" to which they have yet to contribute wastes.

Many argue that creating incentives to manage hazardous substances properly is the job of the RCRA program, the logic being that RCRA is the federal law that sets standards for the treatment, storage, and disposal of hazardous wastes. There are, however, substantive differences in the breadth of liability under RCRA and Superfund. First, RCRA regulates only wastes, whereas Superfund liability attaches to a broader category of hazardous substances. Second, not all wastes that pose a hazard are regulated under RCRA; for instance, wastes from mining and also oil and gas operations are exempt. Thus, even if RCRA were working exactly as intended—a claim not often heard—it would not necessarily promote due care in waste handling across the board. It is our feeling that Superfund fills the void and also acts to reinforce the positive incentives RCRA creates for wastes covered under that law.

Fairness. Superfund has been and will continue to be criticized on grounds of fairness, and not without justification. First, strict liability holds parties responsible for cleanup not only at sites where all then-prevailing rules and regulations were complied with, but also in some cases where PRPs made special efforts to ensure safe disposal. Second, joint and several liability can result in one PRP paying for contamination caused by others, or in one or more PRPs paying for the careless practices of the owner/operators of the sites to which they sent wastes. Both of these latter features of the law are inconsistent with the polluter-pays principle.

Superfund liability is also retroactive and thus inconsistent with the notion that changes in rules should be prospective wherever possible. Retroactivity has been a bitter pill for PRPs to swallow, especially those who had been making special efforts to handle wastes responsibly. And, although they are not directly liable under Superfund, insurers have been adversely affected by its retroactive liability as well. According to the insurers, because they did not anticipate that the policies they issued would sometimes be interpreted to cover nonsudden releases of contamination, they did not charge premiums or set aside reserves to cover these potential liabilities. Thus, retroactive liability greatly increases their exposure and could, in the extreme, materially affect their financial viability. There is one respect in which Superfund accords better with fairness, however: the liability standards generally treat all (solvent) PRPs equally, holding each potentially liable for cleanups at sites for which they have some involvement.

Although it receives less attention, perhaps the greatest inequity in Superfund is a direct result of the insolvency or disappearance of a number of PRPs. When this occurs, either the remaining financially viable PRPs (sometimes referred to as "deep pockets") or the Trust Fund is left paying the bill for these "orphan" PRPs. In addition, many PRPs complain that EPA does not even involve all known and solvent parties in settlement and enforcement actions—a major source of resentment for those who are named.

We present some data here that shed a bit of light on the extent to which Superfund is in rough accord with the notion that those responsible for the problem should pay to clean it up. (See appendix A in this report for details on the survey from which the data are taken.) For instance:

• Of 903 NPL sites for which EPA collected and provided Resources for the Future (RFF) new information on the cause of site contamination, owner/operators were the only

27

parties responsible at 240, or 27 percent, of the sites. Thus, although retroactive liability was clearly an *ex post facto* change in the rules, holding the PRPs at these sites liable for cleanup costs appears not to be terribly unfair, considering that someone has to pay.

- At 89 of the same 903 sites, contamination was caused by operations that were illegal at the time they were carried out. At approximately 35 (39 percent) of these 89 sites, the illegal contamination appears to have been the result of actions by the owner/operator. In these cases, holding those responsible for the contamination would be warranted, even under liability systems where "negligence" is the key determinant of liability.

Thus, at one-third of the sites for which we obtained information, it seems not entirely unfair to hold the responsible parties liable—so long as the costs are falling on the parties that had owned and operated the sites continuously in the first case and on those responsible for the illegal activities in the second. We tried but failed to obtain information on how costs were being assigned at these sites—that is, on how many PRPs were being held liable and whether they were paying their "share." The data we did collect imply, however, that at many sites, liability may *not* be being assigned in a particularly equitable fashion. Thus, evaluating "fairness" in Superfund is no easy task. In some cases, it probably accords reasonably well with common notions of fairness, while in other cases it may not.

As part of the EPA/RFF survey, information was also collected on the number of sites affected by the retroactive nature of Superfund liability (see appendix A). We found the following:

- Of the 708 sites for which EPA could determine the date when waste disposal operations ceased, 369 sites (or 52 percent) had closed these operations before 1981. For these sites, then, rules were changed *ex post facto*. Interestingly, of these approximately 700 sites, 85 (12 percent) are *still* accepting wastes for disposal.
- However, of the 369 sites that closed prior to 1981, contamination was caused by the owner/operators at about 30 percent, while at 15 percent of the sites illegal disposal had taken place. Thus, although liability has been applied retroactively at just over one-half of the sites we surveyed, in about half of those cases it would be difficult to argue that someone other than the PRPs ought to pay for cleanup.

Fairness is easier to evaluate in situations where there are few PRPs or where the PRP was clearly negligent than, for example, at a multiparty site where the owner/operator mishandled wastes years ago and is now insolvent. In this latter situation, making the generators pay for cleanup seems unfair, especially if the waste disposal took place before Superfund became law. That some PRPs are not financially viable is the one fact that makes it difficult for any Superfund financing scheme to be "fair," unless general revenues are used to stock the Trust Fund.

One final observation. In our view, it is unfortunate that Superfund has come to be defended primarily as a polluter-pays statute. CERCLA liability provisions were intended much more as a means of financing cleanups without increasing federal spending than as a means of fingering "guilty" parties. While our data show that in many cases the assignment of financial responsibility under Superfund may not be altogether unfair, in other cases it clearly is.

Financial Implications.   Startlingly little information is available on the actual aggregate costs of site cleanups and the identity of the parties paying for studies and cleanup at individual sites. Over the past 18 months we tried to obtain this kind of information from EPA, the Department of Justice, trade associations, and other organizations. We had no success. EPA does not ask PRPs to report what they actually spend on NPL site studies and cleanups, even when this work is done under some kind of legal agreement with the agency. Neither have trade associations made public any data on cleanup costs, either in the aggregate or by company. What *is* known are the following: how much EPA is spending on cleanups; how much various parties are paying in taxes to finance the Trust Fund; how much the federal government has received in all cost recovery actions to date; and how much in future settlements has been agreed to by PRPs. In addition, EPA has released estimates of the total amount of funds it will need in the future to clean up those sites now on the NPL.

EPA has estimated that it will eventually incur total costs of just over $27 billion for the cleanup of all sites now on the NPL and the implementation of other aspects of the Superfund program.[17] Slightly more than $18 billion of this amount remained to be spent in FY 1992 and beyond. Of the $27 billion, approximately $7 billion is projected to finance the EPA portion of remedial actions (RAs).[18] The remaining $20 billion includes EPA's costs for removal actions and all other aspects of the Superfund program (EPA-lead remedial investigation/feasibility studies and remedial designs, oversight of PRP-lead activities, removal actions, site discovery and listing activities, PRP searches and enforcement, and so on). Annually, EPA's "non-RA" costs are approximately $1.0 billion.

EPA's estimates of its RA costs are based on three key assumptions: (1) an average site cleanup will cost $27 million; (2) EPA will take the lead for 35 percent of future remedial actions; and (3) RA costs are estimated only for the 1,115 sites on the current NPL.[19]

Based on these assumptions, *total* remedial action costs for these NPL sites would be just over $30 billion (1,115 sites x $27 million per site). If EPA pays for $7.3 billion of these costs as the agency has projected, PRPs would have to contribute $22.7 billion—an EPA/PRP split of about 1:3 in actual dollars. Of course, the actual PRP share will rise if EPA successfully recoups some of its RA costs through cost recovery actions. Through the end of FY 1991, the total value of PRP settlements (which will be expended over a number of years) was $3.5 billion; this suggests that few of the dollars ultimately expected to come from PRPs for cleanup have been committed to date.[20]

EPA's estimate of its own total Superfund expenditures of $27 billion is almost certainly an underestimate. First, according to EPA, as of the year 2000 there will still be

[17] U.S. Environmental Protection Agency, Office of Emergency and Remedial Response, *Progress Toward Implementing Superfund, Fiscal Year 1990: Report to Congress* (Washington, D.C., EPA, February 1992), p. 37.

[18] U.S. Environmental Protection Agency, Office of Emergency and Remedial Response, *Progress Toward Implementing Superfund, Fiscal Year 1989: Report to Congress* (Washington, D.C., EPA, 1990), p. 41. No similar figure is provided in the more recent FY 1990 report. Note: annual EPA costs of $752.9 million for "site cleanup" in FY1991, as reported in the FY 1990 report, include funds for RAs *and* for removal actions, site studies, remedial design, and PRP oversight.

[19] As stated in chapter 2, this was the number of nonfederal sites on the NPL as of July 1991, when RFF obtained CERCLIS data from EPA. Although the agency plans to add 100 new sites to the NPL annually, no estimates of the additional funding these sites will require have been made public.

[20] U.S. Environmental Protection Agency, Office of Solid Waste and Emergency Response, *Superfund FY 1991 Annual Management Reports;* data as of September 30, 1991 (Washington, D.C., EPA, 1991).

600 sites where cleanup is not complete. Thus, the agency will incur administrative costs for these sites past the year 2000—possibly for an additional 9 or 10 years. Second, even after implementation of the remedy is completed at all sites, EPA may still need funds to cover groundwater treatment and monitoring at some sites (which can go on for decades). Third, many believe that the average cost of cleanup will greatly exceed $27 million per site. This belief was bolstered by a recent study by researchers at the University of Tennessee, who concluded that cleanup costs are likely to average $50 million for each of the sites now on the NPL.[21] If these new estimates are correct, either EPA or PRPs or both will have to spend more money on cleanups than current projections indicate. It is worth noting, however, that RA costs could also decrease, if advances in new technologies are forthcoming.

As noted in chapter 2, we have elected to use an average site cleanup cost of $40 million for the 1,115 nonfederal sites on the NPL. This implies a total cleanup cost of $44.6 billion. Assuming no increase in EPA's planned RA expenditures of $7.3 billion—which is based on average site cleanup costs of $27 million—PRPs would need to finance $37.3 billion, or 84 percent of the total cost of remedial actions at current NPL sites. Because we assume no changes in liability in Option 1, no increase is required in the corporate environmental tax. Note that we do not estimate EPA's likely future costs for any sites added to the NPL in the future.

Option 2: Expanded Mixed Funding for Orphan Shares

Under Option 2, the Trust Fund would be increased to cover those cleanup costs attributable to parties not participating in EPA settlements. This would include "orphan" shares—that is, shares of parties that are no longer in business or are insolvent, as well as those of recalcitrant parties. In this report we use the term to apply, inclusively, to both, for editorial ease. EPA would continue to bring cost recovery actions against recalcitrants to recover their portion of cleanup and other costs.

Speed of Cleanup.   If orphan shares were fully covered by the Trust Fund, it is likely that negotiations between EPA and PRPs and cost allocation among private parties would be expedited, perhaps considerably. PRPs would not be paying for problems created by others, which should make them more willing to come to the negotiating table quickly. In addition, of course, PRPs' financial contributions to cleanup would be reduced because they would not, as a matter of policy, be held liable for the shares of insolvent or uncooperative parties, which in turn should reduce their incentive to delay cleanups. On the basis of principle and in pragmatic economic terms, assumption by the Trust Fund of orphan shares should translate into a reduction in the time spent before cooperating PRPs enter into agreements with EPA to undertake site study and cleanup activities.

Expanded mixed funding could be difficult to implement, however, and might even delay cleanup in some cases. For example, after nearly 12 years of experience under Superfund, there is still little agreement as to how costs should be allocated: Should it be done on a volumetric basis? According to toxicity? What is the "right" share for owner/

---

[21] E. W. Colglazier, Tracy Cox, and Kim Davis, *Estimating Resource Requirements for NPL Sites* (Knoxville, Tenn., University of Tennessee, Waste Management Research and Education Institute, December 1991).

operators? A recent effort by EPA to develop a model allocation for municipal landfills—and the resulting controversy about whether costs should be allocated based on the volume of waste contributed, the toxicity of the waste, or the relation of the waste to the cost of cleanup—dramatically illustrates just how difficult this can be. If the burden were on EPA to estimate the orphan share, the agency might be required to conduct a nonbinding allocation of responsibility for each site where mixed funding is a possibility. Disputes over the "right" size of the orphan share could delay cleanup in these cases. Moreover, even with expanded mixed funding of orphan shares, PRPs could still be held jointly and severally liable for nonorphan shares *if* EPA did not bring all responsible parties into initial settlement negotiations or enforcement actions. In addition, where PRPs seek to delay paying for cleanup, requesting a mixed funding agreement could be used as a new delaying tactic. We should point out, however, that all these problems exist under the current Superfund program (Option 1).

If expanded mixed funding was to speed cleanup, clear rules would have to be established for the circumstances in which it would be considered, for the apportionment of shares, and for fixing the burden of specifying the size of the orphan share.

Transactions Costs. As noted above, expanded mixed funding would be likely to reduce legal and other costs to PRPs. Cost allocation among PRPs should be made easier, in part because less money would be at issue. Option 2 could, however, shift some transactions costs from the private to the public sector. For example, because the Trust Fund would cover recalcitrants' shares of cleanup, the cost of going after nonsettlors would lie completely with EPA; this might give some PRPs an incentive to be recalcitrant and to wait for EPA to bring cost recovery actions. Perhaps most importantly, as long as the size of the orphan share must be determined, there would be some transactions costs involved in identifying orphan PRPs and assessing what percentage of cleanup costs at a site were their responsibility. Still, these costs are likely to be considerably smaller than those resulting from litigation among PRPs, which is how PRPs reallocate costs after their initial agreement with EPA under the current program.

Option 2 would, however, have little or no effect on two other tiers of transactions costs—those between PRPs and insurers, and those between insurers and re-insurers—unless it greatly reduced the total costs all these parties face. Even if individual PRPs were only paying their "fair share" of costs at a site, in other words, they would want to shift these costs to insurers wherever possible. Similarly, insurers finding themselves holding the bag for cleanup costs would no doubt turn to re-insurers for remuneration.

Voluntary Cleanups at Non-NPL Sites. Expanded mixed funding of orphan shares would, by definition, only affect multiparty sites. We assume that any site that is truly orphan (that is, one for which there exist *no* financially viable or cooperating parties) is already scheduled for cleanup or is being cleaned up with Trust Fund monies. Thus, any PRP that is cleaning up potential Superfund sites on its property under the current program would still have an incentive to do so should the size of the Trust Fund be expanded to pay for orphan shares.

To the extent that companies voluntarily undertake cleanups at multiparty sites to which they have contributed only a portion of the wastes, Option 2 would have a negative effect on voluntary non-NPL cleanups. Expanded mixed funding could actually slow down cleanups at these sites, because the solvent parties would have an incentive under Option 2 to wait for EPA to first assess and then pick up the orphan share.

Due Care in Future Waste Management Practices.   Expanded mixed funding of the shares of insolvents and recalcitrants would not be expected to affect the commitment of either private or public parties to care in waste handling and disposal. Those who are managing hazardous substances with considerable care today would likely continue to do so.

Some companies, for example, are sending their wastes to hazardous waste management facilities in order to minimize future liability, even when there is no regulatory requirement to do so.[22] They would in all probability continue to do so under Option 2, since it would not affect their likely future financial exposure. These companies know that they could be liable as PRPs in the future if the hazardous waste facilities to which they send their wastes become Superfund sites. They also know they would be liable for their share of site remediation no matter where the wastes were deposited. Their only escape from financial responsibility would be bankruptcy, and this "incentive" exists under the current program. Those PRPs that have not improved their waste management practices or attempted to reduce their use of toxic substances would not be affected by Option 2.

Fairness.   Aggressive mixed funding would redress one of the most obvious inequities in the current system—having one PRP pay more than its "share" of cleanup costs. In this sense, then, it is in keeping with the spirit that those responsible for the problem should pay—that is, with the polluter-pays principle.

Of course, the extent to which expanded use of mixed funding would be more equitable than the current program depends in large part on where the additional monies needed to augment the Trust Fund would come from. In this report, we posit that increased revenues would come from an increase in the corporate environmental tax. Because Trust Fund monies would be covering orphan shares under Option 2, it is axiomatic that some of the companies that "should" be paying are not; insolvent or defunct companies obviously will not be paying increased Trust Fund taxes no matter how these taxes are levied. As long as EPA pursued cost recovery from recalcitrants, no similar problem of equity would arise. If those paying increased taxes were roughly the same companies that would pay under the current program (Option 1), this option would be no more unfair than the current system. If Option 2 also reduced these same companies' transactions costs, however, it might well fare better on the fairness scale.

Financial Implications.   We assume that true orphan sites are already being paid for by the Trust Fund under the current program. Thus, the increased costs to the Trust Fund associated with Option 2 would result from covering orphan shares at sites where there is a mix of responsible parties—some financially viable or cooperative, others not. It is commonly suggested that the percentage of orphan sites on the NPL is 20 percent; recent data collected by EPA for Resources for the Future indicate that "true" orphan sites may represent closer to 12 percent of the sites on the NPL (see appendix A in this report). There are no data, however, on how many of the other sites have orphan PRPs or what percentage of site cleanup costs these orphan shares represent.

---

[22] Department of Natural Resources, Bureau of Research, *Reducing Hazardous Waste in Wisconsin: A Summary of the Barriers and Incentives Wisconsin's Firms Encounter* (Wisconsin, March 1991).

The EPA data collected for Resources for the Future suggest that, at a minimum, 27 percent of NPL sites were contaminated solely by the owner/operator; that is, they are single-party sites. Thus, at most, 73 percent of current NPL sites could be multiparty and therefore could have an orphan share. Lacking any reliable information on what the aggregate orphan share is at multiparty sites, we hypothesize 10 percent. Thus, under Option 2 the Trust Fund would absorb 10 percent of the remedial action costs at 73 percent of the 1,115 sites now on the NPL. This would be an increase of $325.6 million on a yearly basis, or a 71 percent increase in the corporate environmental tax. (See appendix C for an explanation of how these figures were developed.) Again, we do not estimate any additional EPA costs under this option for sites added to the NPL in the future.

Option 3: Liability Release for All Closed Co-disposal Facilities

To our knowledge, there are no reliable data on the number of NPL sites at which municipal wastes and industrial wastes have been co-disposed, nor the number of these sites which have stopped accepting wastes for disposal, that is, which have closed. Some information exists on how many NPL sites are municipal solid waste landfills (that is, *not* hazardous waste landfills regulated under Subtitle C of RCRA) and how many of these are publicly and privately owned.

We summarize here the best available data on municipal landfills—those sites most likely to be co-disposal sites.

- According to EPA's CERCLIS database, there are 237 landfills on the NPL.
- According to information developed by EPA's Office of Waste Programs Enforcement, there are 214 NPL sites where municipal solid waste was deposited.[23] All but 70 of these sites are also listed as landfills in CERCLIS.
- According to the EPA/RFF survey, 174 (or 19 percent) of the 899 sites for which data were provided are owned and/or operated by municipalities (see appendix A).
- According to a recent report by Clean Sites, Inc., 179 sites on the NPL are owned by local governments, 11 NPL sites were previously owned by local governments, and 148 privately owned NPL sites have seen some local government involvement, although no information is provided as to whether these local government PRPs were operators, generators, or transporters.[24]

Unfortunately, these data are of limited assistance in estimating the number of sites on the NPL where municipal wastes and industrial wastes were co-disposed—sites that present the most controversial cost allocation issues.[25] Based loosely on the data reviewed

---

[23] Personal communication from Arthur Weissman, Office of Waste Programs Enforcement, U.S. Environmental Protection Agency, 1991.

[24] Clean Sites, *Main Street Meets Superfund: Local Government Involvement at Superfund Hazardous Waste Sites* (Alexandria, Va., January 1992).

[25] It should be noted that any waste management facility that accepted waste or products for treatment, disposal, or recycling from many generators—whether municipal solid wastes alone or hazardous wastes—also presents difficult allocation issues.

above, and lacking any better basis, we assume that 250 sites on the current NPL are co-disposal facilities that are closed, that is, no longer accepting wastes; we use this estimate in our calculations of the financial implications of this option.

Speed of Cleanup.   Releasing both local governments and all private parties from liability for the cleanup of all 250 NPL sites at which municipal and industrial wastes were commingled would surely accelerate cleanup at these sites. This follows directly from the elimination of the need for settlement negotiations between EPA and PRPs and also for cost allocation among PRPs. Most importantly, perhaps, it would speed cleanup because Trust Fund monies would be available immediately to commence site study and cleanup activities.

Transactions Costs.   Eliminating liability where co-disposal has taken place would eliminate much of the intra-PRP litigation at sites many people believe to be the most contentious. It would also eliminate any PRP-insurer or insurer–re-insurer litigation resulting from Superfund liability at these sites, since no PRPs would be held liable. Thus, familiar kinds of transactions costs would be eliminated for almost 25 percent of NPL sites if this option were implemented. If cleanup costs average $40 million per site, and if transactions costs average 20 percent of cleanup costs (a figure we choose somewhat arbitrarily), Option 3 could save $2 billion in transactions costs. We must add two caveats, however. First, this estimate is extremely uncertain; it is only as good as the assumptions on which it is based. If it is in the right ballpark, however, it suggests considerable savings, since $2 billion is enough to finance an additional 50 cleanups at our assumed per site cost. Second, these savings would not accrue all at once, of course. If spread over 10 years, the savings amount to $200 million (or five additional cleanups) per year.

If all PRPs at closed co-disposal sites were released from liability, some new transactions costs would be created. As noted in chapter 2, however, most of these new costs could be addressed legislatively. For instance, Congress could elect not to reimburse PRPs for site study and cleanup costs already incurred at sites where co-disposal of wastes took place. Similarly, liens could be placed on any property cleaned up with Trust Fund monies. But since the trigger for a release from current Superfund liability under Option 3 is co-disposal, both municipalities and private firms would have a strong incentive to demonstrate that co-disposal took place at sites where they are PRPs—even in cases where that assertion would be open to question. Since the financial stakes would be quite high, these parties could find it worthwhile to engage in extensive litigation over this issue. Thus, very clear definitions of what constitutes co-disposal and what constitutes a closed site would be necessary.

Voluntary Cleanups at Non-NPL Sites.   Option 3 would apply only to sites that were closed before the date on which CERCLA is amended. Thus, legislative deliberations over such a change could lead to a slowdown in cleanups for such sites, as the PRPs involved waited to see if Congress would take them off the hook. It could also create an incentive for similar facilities to close in order to obtain Trust Fund-financed remediation. The number of such facilities is likely to be small. For instance, EPA/RFF survey data suggest that of those 174 NPL sites owned by municipalities, 22 are still operating.

Since Option 3 affects only multiparty sites, it would not blunt any existing incentive for PRPs to clean up contamination at non-NPL sites for which they are the sole PRP.

Option 3 would, however, reduce any incentive to clean up multiparty sites cooperatively when these are closed co-disposal facilities.

Due Care in Future Waste Management Practices. Eliminating liability for closed co-disposal sites should have little effect on existing incentives *unless:* (1) PRPs believe it would be possible to expand this liability release to similar sites that cease operation in the future; or (2) they believe other relaxations of the law will be forthcoming. If PRPs thought either or both of these to be the case, it could reduce current incentives for owners/operators to handle wastes carefully (knowing they could get out of responsibility for cleanup) and would create an incentive for generators to ship wastes to co-disposal facilities. A likely result would be more and more substances disposed of off site and little incentive to assure good off-site management of hazardous substances.

Fairness. While it would go some way to redressing inequities created by retroactivity (for those closed co-disposal sites contaminated by pre-1981 activities), a liability release for closed co-disposal sites would interject a new inequity into the Superfund law. Specifically, PRPs who were lucky enough to co-dispose of their wastes before 1994 at closed facilities (or whenever Superfund is next amended) would be released from Superfund liability, whereas those that managed identical wastes either on-site or in landfills where industrial and municipal wastes were not commingled would still be subject to Superfund liability. Thus, liability would depend on where wastes happen to have been managed rather than on their inherent toxicity or on the risks they engender. In addition, in some ways this option is quite unfair in that it releases some PRPs from Superfund liability for actions they took *after* Superfund became law. In other words, apparent equals would not be receiving equal treatment. Finally, as noted earlier, fairness would depend in large part on who is paying increased taxes to finance the additional Trust Fund revenues.

Financial Implications. In estimating the additional cost to the Trust Fund, we assume that few of the 250 closed co-disposal sites that would be released from liability under Option 3 are being cleaned up now as orphan sites. Thus, all monies to pay for cleanup of these 250 sites would need to come from additional Trust Fund monies. We again use $40 million per site as an average cleanup cost.[26]

To spend $40 million per site for an additional 250 sites, the Trust Fund would need an additional $10 billion over 10 years, or $1 billion annually for those closed co-disposal sites currently on the NPL. This would require a 217 percent increase in the corporate environmental tax. (See appendix C for an explanation of how costs were developed.) Note that this is almost certainly an underestimate of the Trust Fund implications for Option 3, as we do not estimate how many of the sites added to the NPL in the future are likely to be closed co-disposal sites.

---

[26] Although horror stories abound about landfills where total cleanup costs could eventually range into the hundreds of millions of dollars, careful analysis suggests the costs to date of cleaning up landfills on the NPL may not be much greater than for other sites. Estimates by researchers at the University of Tennessee [Colglazier, Cox, and Davis, *Estimating Resource Requirements for NPL Sites*] suggest an average landfill remediation cost of $40 million as compared with an average site cleanup cost for all NPL sites of $50 million. EPA's own records suggest a slightly higher than average remediation cost for landfills: $34 million ($15 million per operable unit with an average of 2.25 operable units per site) versus the $27-million cleanup figure EPA uses for all sites.

More than any other option, Option 3 calls into question our assumption that all additional monies for the Trust Fund would come from an increase in the corporate environmental tax and cries out for a more broadly based financing mechanism. For although corporate PRPs would benefit from a liability release for co-disposal sites, so too would municipalities and smaller companies that do not pay the corporate environmental tax. Yet the latter would contribute nothing toward the increased Trust Fund and would therefore get a "free ride" under this approach. In a different economic (and political) climate, one would be tempted to fund a portion of the increased revenues needed from the federal budget.

Option 4: Liability Release for All Pre-1981 Sites

Under Option 4, the Trust Fund would pay for cleanup of all NPL sites where waste disposal operations ceased before January 1, 1981, which we choose as the cut-off date because it conveniently approximates the date (in December 1980) when CERCLA became law.

An issue crucial to Option 4 is specifying to which activities the cut-off date applies: it could be applied PRP-by-PRP, based on when each PRP last sent hazardous substances to a site; it could be applied on a site-by-site basis, in which case site closure could be determined by the date when hazardous substances were last sent to a site, the date on which the actual waste disposal operation (landfill, surface impoundment, and so on) at the facility closed, or the date when the facility (for example, the chemical or wood-preserving plant) at which the site was located closed. At some Superfund sites there are no "facilities" per se. In these cases, the cut-off date would have to be determined either on a PRP-by-PRP basis or when the contamination occurred (that is, when hazardous substances were "deposited" at the site). Each of these definitions would have important implications for the number of sites to be cleaned up with Trust Fund monies, as well as for the administrative feasibility and relative fairness of the option.

Arguably, a cut-off date based on when individual PRPs last sent hazardous substances to a site would be the fairest method of defining which PRPs would retain liability. (The logic is that the passage of Superfund in December 1980 signaled a new era in waste management to which all responsible parties should have been sensitive, and thus substances sent before that time should escape liability.) However, we are concerned that this definition would be difficult to implement administratively because it would require establishing definitively who sent which substances to a site and when. This could perhaps be addressed by fixing the burden of proof firmly on the PRP to demonstrate that it had "beat" the deadline, although EPA would still have to evaluate the legitimacy of this proof. We believe that the administrative costs of making such determinations could rival those of the current program. For this reason, in Option 4 we define the cut-off date on a site basis and select the date on which the disposal operation closed.[27]

_____

[27] In the recent EPA/RFF survey of EPA regional staff to obtain information on when NPL sites "closed" (see appendix A), we found that at 60 percent of the sites, regional staff said they did not know when hazardous wastes were last deposited at the site. At just over 70 percent of the sites, however, regional EPA staff were able to provide information on the year the facility had closed and the year the waste disposal operation had closed.

Speed of Cleanup.   Releasing all pre-1981 sites from Superfund liability clearly should accelerate cleanup at those sites, since it would eliminate the time-consuming negotiation, enforcement, and settlement processes and would assure that funds would be readily available to initiate cleanups. According to the recent EPA/RFF survey, waste disposal operations ceased before 1981 at 369 (or 52 percent) of the 708 NPL sites for which data on closing were available (see appendix A for survey data). Applying this percentage to all 1,115 NPL sites means that 580 sites may fall within our definition of pre-1981 sites. We use this estimated number throughout our evaluation of Option 4. Thus, even more sites would be affected under Option 4 than under Option 3.

One cannot rule out the possibility that enactment of a legislative change similar to the Option 4 approach could *slow* cleanup, at least at some sites. This might happen for sites where there are protracted disputes over the date on which the disposal operation ceased. In addition, since many sites have multiple sources of contamination, defining when disposal operations ceased could be complicated. At sites with multiple landfill units, for example, some may have closed before 1981 and some after; would these sites be treated as "pre-1981" or not? The resolution of such questions could conceivably slow cleanup at some sites, although the number of sites affected by this issue would probably not be very large. These issues could—and should—be addressed in legislative language if a revised liability scheme similar to Option 4 were to be implemented.

Transactions Costs.   Under Option 4, at those sites now to be cleaned up by the Trust Fund, negotiations between PRPs and EPA as well as third-party litigation would be eliminated, as would related litigation over insurance coverage. At $40 million per cleanup, and with 580 sites, if transactions costs are 20 percent of site remediation, transactions cost savings of about $4.6 billion—over some number of years—could be realized under Option 4. Again, however, we caution readers as to the very speculative nature of this estimate.

One new kind of transactions cost would be likely to develop under this option. As suggested above, litigation about when a disposal operation closed would be likely to arise at some sites. No matter how well new legislation might define what it means to be an "old" versus a "new" site, it is almost certain that this definition would be litigated aggressively. To give some indication of the number of sites that might be affected, we look to information collected in the recent EPA/RFF survey on when NPL sites closed their disposal operations. Based on this information, it appears that 108 NPL sites (or 14 percent of those sites for which we obtained this information) closed in 1981 through 1983. These are sites for which litigation is most likely. Being as explicit as possible in defining where liability has been eliminated would be crucial to keeping this new source of litigation to a minimum. Even if such transactions costs do arise, however, we are quite confident they would not begin to approach the savings likely to accrue from reduced legal costs at other sites.

To summarize, transactions costs under Option 4 should be greatly reduced at pre-1981 sites as multiple tiers of litigation disappeared. Some new costs could be incurred, however, where there was ambiguity about the date on which waste management operations ceased. The net effect of these changes on transactions costs would depend to some extent, however, on the characteristics of pre-1981 sites. If most pre-1981 sites involved a number of PRPs, this option could greatly decrease transactions costs. If most pre-1981 sites are single-party, however, the effect is less clear; the recently released RAND report suggests a large discrepancy between transactions costs at single-party and multiparty sites.[28] Our data suggest a mix of pre-1981 sites—which means that transactions costs would be reduced significantly for some, but not all, of these sites. Any more sophisticated analysis of the likely effect of changes in liability would require more and better data on transactions costs and better information on the mix of pre-1981 sites on the NPL.

Voluntary Cleanups at Non-NPL Sites. Eliminating Superfund liability for pre-1981 sites would greatly reduce the incentive for the voluntary cleanup of any such sites not yet on the NPL, except in states that require cleanup for commercial property transactions. Suppose, for example, that a company owns property on which is located a surface impoundment that clearly closed in 1978. If the company wished to sell that property, neither it nor the prospective purchaser would have much of an incentive under Option 4 to have the contamination cleaned up as a condition of the sale (unless the property is located at a facility that has a permit under Subtitle C of RCRA), since neither would be held liable under Superfund for site cleanup under this option. The only motivation that would exist for a company to conduct cleanups of pre-1981, non-NPL sites would be liability under state Superfund laws, toxic tort liability, or a corporate policy requiring such cleanup. While these are by no means trivial incentives, Superfund liability adds to their force.

In principle, a modification to Superfund's liability standards along the lines of Option 4 should not be expected to affect the incentive for cleanup at sites not on the NPL that closed after December 31, 1980. This is because Superfund liability would continue to attach to these sites under this option. In practice, this incentive might be diminished somewhat if responsible parties believed they might be able to extend the liability exemption to even more recent sites, or if site records are ambiguous about when the waste management facility ceased operation.

Due Care in Future Waste Management Practices. It seems reasonable to infer that eliminating retroactive liability for pre-1981 sites should have no effect on future waste management practices. After all, why should decisions about the treatment of bygone problems (the lax management of hazardous substances) affect future practices, especially in light of a welter of other current laws influencing the latter? Nevertheless, we believe the issue to be more complicated than this. In our view, the imposition of retroactive liability in 1980 focused the attention of those in the public and private sectors on the costs of improper management of hazardous substances. As a result, we believe that Superfund liability has been one of the factors that has prompted increased efforts at waste reduction and more careful management of hazardous substances. Depending on other sanctions governing waste management practices, eliminating liability for pre-1981 sites now might send the wrong signal to both government and private PRPs and reduce their incentives for sound *prospective* management of hazardous substances.

Fairness. Assumption by the Trust Fund of all cleanup costs for pre-1981 sites addresses the inequity that many saw (and still see) in the imposition of retroactive liability in 1980. Nevertheless, it represents a kind of "rough justice" to address that inequity. This is because

[28] Acton and others, *Superfund and Transaction Costs,* pp. 51-53.

38

we have defined Option 4 as pertaining to those sites at which waste disposal operations closed prior to 1981. Thus, PRPs that were behaving responsibly as generators, transporters, or disposers of hazardous substances but that had the misfortune to send substances to sites whose waste disposal operations ceased after December 31, 1980, would be held liable for a higher standard of care than those using sites that closed prior to the cut-off date. In this respect, not all "polluters"—even almost identical ones—would be made to pay equally. Thus, while Option 4 has appeal on grounds of fairness because some aspects of retroactive liability are eliminated, it would make an unfair distinction among PRPs based solely on when waste disposal operations ceased.

Congress could, of course, address this concern by defining "pre-1981" as the date before which hazardous substances were *sent* to a site. Although this definition of the cut-off date is much more appealing on grounds of fairness, it has very considerable practical implications. First, it would require the same need that exists today to identify PRPs and link them to the disposal of specific substances. Second, it is conceivable that *all* NPL sites now and in the future could have *some* pre-1981 wastes, meaning that the liability release would apply to some portion of all future NPL sites. Such a system could result in both liability-based and Trust Fund-financed cleanups going on simultaneously at most NPL sites, as would happen under Option 2. This could prove bureaucratically cumbersome.

In addition, some pre-1981 sites would inevitably be single-party sites, and assumption by the Trust Fund of cleanup costs at such sites would be a clear violation of the notion that those responsible for contamination should pay for cleanup (although admittedly in a case where the "polluter" may have been following the best practices of the time). To put the matter somewhat differently, one might ask why taxes on chemical and petroleum feedstocks and a broadly based corporate environmental tax should be used to clean up a site for which the owner is both identifiable and solvent (and in some cases, at least, may pay relatively little in these taxes). Some have suggested that Congress simply exempt single-party sites from any liability release to address this concern. Others point out that the implication of such a carve-out would be that it is acceptable to hold PRPs liable at single-party sites, but not at multiparty sites—which may be more inequitable than the imposition of retroactive liability. This aspect of Option 4 then, makes it more difficult to judge it a complete success on grounds of fairness.

Financial Implications.   To estimate the additional resources that the Trust Fund would require under Option 4, we first assume that none of the 580 pre-1981 sites is truly an orphan site, that is, a site where there are no financially viable PRPs and all the costs of the site are already the responsibility of the Trust Fund. Thus, under this option all 580 sites would revert to the Trust Fund for cleanup, adding a total of $23.2 billion over 10 years to EPA's costs, or $2.3 billion annually. This in turn implies an increase of 500 percent in the corporate environmental tax (see appendix C for an explanation of how the cost estimates are derived). Again, this estimate of needed additional Trust Fund monies does not include costs to EPA of pre-1981 sites added to the NPL in the future.

Option 5: Liability Release for Current NPL Sites

Option 5 would transfer cleanup responsibility to the Trust Fund for all the sites on the NPL on the date Superfund is amended. Sites included on the NPL after that date would be cleaned up according to the liability standards of the current program (Option 1).

Speed of Cleanup.   On its face, it would appear that eliminating PRP liability for all sites on the NPL would speed cleanups significantly. After all, it would eliminate virtually all the delay related to negotiation or litigation under the Superfund program. Moreover, by assumption, the monies for site studies and cleanups would be readily available from the augmented Trust Fund. As a result, Option 5 could be expected to hasten the cleanup process significantly. However, there would be no change in liability from the current program for sites added to the NPL after the law was amended.

Several factors could mitigate such a dramatic acceleration of cleanup. First, under Option 5, EPA would be pressured to clean up 1,115 sites (or however many would be on the NPL) all at one time, something it simply cannot do. Even with ample funds and staff, it is unlikely that the EPA could simultaneously implement cleanups at more than 1,000 sites. A full-scale change in the program, such as would be implied by this option, would probably require development of a new bureaucratic infrastructure at the agency. For example, merely getting adequate contracting support to conduct site studies and cleanups at all NPL sites would be a major challenge. In addition, the dynamic that now exists under the current program between PRPs and EPA in the cleanup process (with each keeping an eye on the other) may well lead to more cost-effective and sometimes quicker cleanups, because the best technical analyses both sides can muster are joined in the debate; this dynamic would not exist under Option 5.

Transactions Costs.   Under Option 5, litigation between the government, PRPs, and insurers would be eliminated for all sites on the NPL at the time the law was changed. This would result in an enormous reduction in transactions costs—indeed, that is the primary reason for considering such a dramatic change in the liability standards. For just the 1,115 nonfederal sites on the NPL now, transactions cost savings could amount to $8 billion, under the standard assumptions we have been making ($40-million cleanup per site, 20 percent transactions costs, and 90 percent of sites not orphaned).

Again, however, Option 5 affects only the current NPL. Since the NPL will—according to EPA projections—nearly double in the next 10 years, substantial transactions costs will be incurred at these new sites in the future.

Voluntary Cleanups at Non-NPL Sites.   As long as PRPs believed that no additional liability releases would be forthcoming, Option 5 would have no effect on voluntary non-NPL cleanups after Superfund was amended. However, if PRPs believed that they would not be held liable for cleanup costs at future NPL sites, this would have a chilling effect on non-NPL cleanups.

Due Care in Future Waste Management Practices.   Abolishing liability for all sites on the NPL at the time Superfund is amended should have no effect on any incentive the law may currently provide to improve management of hazardous substances—again, as long

40

as PRPs had no doubts that they would be held liable for any sites added to the NPL in the future. Option 5 would not release PRPs from liability for any sites not on the NPL as of the date the law is amended and thus would not change any PRPs' incentives in their on-going management of hazardous substances.

Fairness. In our view, transferring to the Trust Fund all cleanup costs for sites on the NPL at the time of reauthorization cannot be justified on grounds of fairness alone. Although Option 5 would eliminate inequities created by the imposition of retroactive liability, these could be addressed more "surgically" by Option 4. In fact, in many cases, Option 5 releases PRPs from liability for activities that took place after Superfund was enacted.

Remediating all current NPL sites using the Trust Fund would mean that at least 339 sites at which waste management operations continued even after the passage of CERCLA (85 of which are still in operation; see appendix A) would be remediated at no *direct* expense to the PRPs. (Of course, many would pay in one way or another because of the feedstock taxes and/or the corporate environmental tax.) This would strike many as being unfair and as a violation of the polluter-pays principle, particularly at single-party sites where only one PRP deposited hazardous substances at any one time prior to the amendments implementing Option 5. Furthermore, if such a modification to Superfund's liability provisions were actively debated, it is likely that the 85 sites with currently operating waste disposal operations would have an incentive to close down these operations quickly so as to qualify for cleanup financed by the Trust Fund.

Financial Implications. The Trust Fund would need to be increased to a total of just under $5.3 billion annually to pay for the cleanup of all 1,115 nonfederal facility sites now on the NPL, under our standard assumption that cleanup costs would be spread over 10 years (see appendix C). As a result, the corporate environmental tax would need to raise $4.2 billion yearly, an increase of 800 percent. This is an underestimate of needed Trust Fund revenues, as some new sites will probably be added to the NPL between now and when Superfund is amended. (See appendix C for explanation of how cost estimates were developed.) Annual Trust Fund levels could be lower if—because of the burden placed on EPA by this option—Congress decided not to have EPA work on all 1,000-plus sites simultaneously.

Summary of Major Conclusions

Table 1 (on page 42) summarizes the major conclusions presented in this chapter regarding the effects of each option. (See appendix D for implications of each policy option on the distribution of Trust Fund revenues.)

41

Table 1. Comparison of Five Superfund Liability Options

Options	Speed of Cleanup	Transactions Costs	Non-NPL Cleanups	Due Care in Future Waste Mgmt. Practices	Fairness	Financial Implications
				Evaluative Criteria		
1. The Status Quo	Slow; delays caused both by protracted negotiation/litigation and by EPA's implementation strategy	Significant; by educated guess—20 percent of cleanup cost; very uncertain	Uncertain; but positive effect	Prospective liability clearly creates incentive; retroactive liability may reinforce this	Retroactivity unfair in many cases, sometimes not; joint and several liability departs from "polluter pays"	Not applicable
2. Expanded Mixed Funding for Orphan Shares	Should speed cleanups at sites with orphan shares; may not in some cases	Reduced PRP-EPA and intra-PRP litigation; little effect on insurance-related litigation	No perceptible impact on single-party sites; possible negative impact on multi-party sites	No effect	Much more in accord with "polluter-pays"; still would impose retroactive liability	Would require $325.6 million increase per year in Trust Fund (10% of RA costs at 73% of sites)
3. Liability Release for All Closed Co-disposal Facilities	Should speed cleanup at closed co-disposal sites	Largely eliminated for co-disposal sites	Fewer, to the extent that non-NPL closed co-disposal sites are being cleaned up	Should have no effect unless PRPs believe future relaxations are likely	Unfair in that liability is determined by where waste was sent	Would require $1.0 billion increase per year in Trust Fund (RA costs for 250 sites)
4. Liability Release for All Pre-1981 Sites	Would expedite cleanup at pre-1981 sites; could delay cleanup at some post-1981 sites	Eliminated at pre-1981 sites; some new costs at other sites	Fewer cleanups at pre-1981 sites	Should have no effect on future actions, although elimination of retroactive liability may give wrong signal	Would eliminate retroactivity; would distinguish unfairly among PRPs based on where waste was sent, not on their practices	Would require $2.3 billion increase per year in Trust Fund (RA costs for 580 sites)
5. Liability Release for Current NPL Sites	Would accelerate cleanups at all sites if EPA has resources to handle greatly expanded program	Eliminated for current NPL; no change for new NPL sites	No effect	Might blunt incentives, since some currently operating facilities would be released from liability	Unfair in letting PRPs at current NPL sites off the hook, but PRPs at new NPL sites would still be liable	Would require $3.7 billion increase per year in Trust Fund (remaining RA costs for 1,115 sites)

4

Conclusions

In this final chapter we step back from the detailed analysis of options in chapter 3 to make several general observations about the existing liability standards in Superfund and the alternatives to them that we have considered.

The next reauthorization provides an opportunity to debate in an open and comprehensive fashion whether fundamental changes should be made in any facets of the Comprehensive Environmental Response, Compensation, and Liability Act (better known as Superfund), including its liability standards. That opportunity should not be missed.

As recounted in chapter 1, Superfund was rushed into law without the kind of protracted debate that often accompanies major environmental statutes. It was in large part a response to the single, highly publicized episode at Love Canal that raised the spectre of widespread and health-threatening contamination in communities across the country. In addition, an impending change of the political guard at the time reinforced in the minds of many legislators the need for quick action. Although the 1986 amendments to Superfund were preceded by substantial debate, virtually none of it dealt with the liability standards that are the subject of this report.

We now have more than 11 years of experience under Superfund, and high hopes for a quick and effective solution to the problem of site cleanup have gradually given way to frustration and cynicism. Opinion polls suggest that the public is still extremely concerned about hazardous waste sites. But those who follow Superfund closely have come increasingly to recognize that cleanups take a very long time to be completed, that resources which contribute nothing to reduced risks to human health and the environment are consumed in the process, and that the program Congress put in place more than a decade ago creates several inequities. Less obvious to all, perhaps, are the salutary effects Superfund may be having on the actions of all those who deal with hazardous substances, not to mention the risk reductions arising from the emergency removals and remedial actions themselves. In other words, Superfund—and particularly its liability provisions—have both positive and negative effects. Similarly, the alternatives to Superfund and its liability standards— including the options described in the preceding chapters—have their own strengths and weaknesses, the most prominent of which we hope have been identified here.

We mention these perhaps obvious points because when we began work on this project 18 months ago, many important players in the Superfund debate seemed unwilling even to discuss possible alternatives to the existing liability and taxing provisions in the law. Such a stance, were it to continue, would be inconsistent with sound environmental policymaking. For while it may have been premature to change directions in the nation's approach to hazardous waste site remediation in the 1986 Superfund reauthorization debate, there is nothing untoward about considering such a change now. Indeed, this is a most appropriate time for stocktaking.

Ultimately, of course, Congress must determine whether any of the options considered here (or others) would, on balance, be better for the country than the current Superfund program. If we are to continue along the Superfund path Congress charted in 1980, that should be the result of an affirmative choice rather than a refusal to consider other possibilities.

The site cleanup process could be expedited by changing Superfund's liability standards. It could also be expedited under the current system.

All would agree that the remedial process takes too long. The U.S. Environmental Protection Agency's own data suggest that it takes nearly 11 years, on average, between the initiation of a site study and the implementation of the remedy at a site on the National Priorities List. While such delays are due in part to the nature of the site study, remedy design, and cleanup processes—features that would be a part of any Superfund program—they are also the result of time-consuming negotiation and litigation between EPA and potentially responsible parties. Releasing PRPs from some or all liability at a subset of NPL sites (as in Options 2 through 5) should reduce these latter sorts of delays.

Just as clearly, however, the remedial process could be accelerated under the existing liability standards. One of the ways this might be accomplished would be for EPA to go back to a "clean up first/recover costs later" strategy that would be roughly consistent with the "polluter pays" principle so long as cost recovery is aggressively pursued. This strategy would require increased revenues for the Trust Fund, but so would all of the alternative liability approaches considered in this report. Second, the current law provides EPA with tools that could speed settlements at sites—such as *de minimis* buyouts, mixed funding, and nonbinding allocations of responsibility—although the promise of these tools is uncertain because they have been little used. Finally, many opportunities exist for EPA to streamline the site evaluation and remedy selection process (as the agency recently proposed), which now takes years.

Transactions costs could be reduced through modification of Superfund liability.

Eliminating the need for EPA to reach agreement on cleanups with PRPs at some sites by releasing them from liability (under Options 3 through 5) would reduce transactions costs. About this we are confident. There is, however, far too little information available on the current magnitude of transactions costs to know just how large the savings would be under each option. We speculate in chapter 3 that the savings would range from $2 billion to $8 billion, albeit spread out over a number of years. Better data are needed on the magnitude of transactions costs at different kinds of sites as well as on the distribution of these sites among all those on the NPL before one could have much confidence in this "guesstimate" of the savings from revising the current liability standards.

44

None of the options we evaluate would eliminate *all* transactions costs at NPL sites, because none would eliminate liability for all sites on the NPL—current and future. Even Option 5—Liability Release for Current NPL Sites—would restrict the liability release to sites on the NPL at the time of reauthorization, but not those added after that date. Also, at least some new litigation would be stimulated under Options 2 through 4 as PRPs sought to show that the sites at which they are involved belong in the category of sites being released from liability. This would be true for any alternative that abolished liability for a subset of sites on the NPL. While these "new" transactions costs may not be trivial, it is inconceivable that they could rival those resulting from the rounds of litigation taking place under the current Superfund liability standards. We simply note that no revision of the present liability standards will eliminate *all* transactions costs. Finally, it is important to note that some alternative approaches that reduce transactions costs for PRPs, such as Option 2—Expanded Mixed Funding for Orphan Shares—do not necessarily reduce transactions costs arising from insurance-coverage disputes.

Relaxing Superfund's liability standards would have some adverse effects.

The most vexing and controversial issue addressed in this report is whether relaxation of Superfund liability (in one form or another) would blunt the incentives many believe the law provides for PRPs both to clean up sites not yet on the NPL and also to manage hazardous substances more carefully. It is important to reiterate that we can provide no convincing statistical evidence that Superfund is having these salutary effects, although recent studies and anecdotal evidence suggest that the effect is a real one.[1] These reports accord well with our own intuition: it simply makes sense to us that PRPs will take measures to keep sites not yet on the NPL off the list. By doing so, they can avoid the sometimes very high costs, protracted delays, and negative publicity that can result from an NPL listing. Any alternative that eliminates liability for a subset of sites could greatly diminish—if not eliminate—this incentive.

It is much more difficult to determine how a change in liability would affect incentives for the careful management of hazardous substances. There is broad agreement that *prospective* liability provides such an incentive. A more difficult question is the effect that retroactive liability under Superfund may have. Some preliminary data support the contention that Superfund does affect firms' waste management practices, especially with respect to substances not currently regulated under the Resource Conservation and Recovery Act (RCRA). Retroactive liability certainly may not be the best way to achieve this objective; indeed, we would be surprised if it was. Nevertheless, one cannot overlook the positive incentive it provides for the careful management of hazardous substances.

On the subject of incentives, another important consideration should not go unmentioned. Those PRPs who have already stepped forward and begun to pay for cleanups at sites like those to be released from liability under Options 2 through 5 would be unfairly treated if, as we suspect is likely, they are not reimbursed for the costs they have incurred. This potential inequity would carry with it a very important incentive effect: the message

---

[1]Jan Paul Acton, Lloyd S. Dixon, and others, *Superfund and Transaction Costs: The Experiences of Insurers and Very Large Industrial Firms* (Santa Monica, Calif., RAND, The Institute for Civil Justice, 1992); Department of Natural Resources, Bureau of Research, *Reducing Hazardous Waste in Wisconsin: A Summary of the Barriers and Incentives Wisconsin's Firms Encounter* (Wisconsin, March 1991).

that it may not be wise to comply immediately with environmental laws because, if one hangs back and waits, the law may be changed. In principle, we find this a very disturbing possibility; it may not be a serious problem in practice given other incentives for compliance and given the relatively small expenditures PRPs have made to date on cleanups.[2]

Any of the modifications to the present liability standards will create at least some new inequities, even as they ameliorate others.

We have identified three definitions of fairness that could reasonably be used to evaluate the Superfund program—first, that polluters should pay for the problems they create; second, that PRPs who handled hazardous substances in a similar fashion should be treated equally; and third, that changes in rules should not be made retroactively.

To the extent that the existing liability rules in Superfund require some "polluters"— that is, solvent PRPs—to pay for problems created by others (insolvent PRPs), they do not comport well with the polluter-pays notion. On the other hand, recently collected data show that at many sites on the NPL, contamination was caused solely by a single owner/ operator. Thus, liability at these sites will attach to the party (or parties) which most observers would view as being responsible for the contamination—whether willful or not. So long as it is possible to ascertain PRPs' "fair shares" at NPL sites, aggressive mixed funding of orphan shares (Option 2) would be the best approach from the standpoint of the polluter-pays principle. Correspondingly, abolishing liability for all NPL sites at the time of reauthorization would probably move farthest from adherence to this principle. Treating similarly situated PRPs equitably is difficult. Options 1 and 2—the Status Quo and Expanded Mixed Funding for Orphan Shares—comport reasonably well with this definition of fairness.

The current program clearly flunks against a standard that holds that any changes in the law be forward-looking only. That is, under Superfund, responsible parties are being held liable retroactively in cases where they followed all the rules in effect at the time (and in some cases where they made special efforts to find safe places to dispose of hazardous substances). If redressing such inequities was deemed to be of paramount importance, then eliminating PRP liability at sites that closed before Superfund was enacted (Option 4) would be especially attractive.

Much better data are needed to assess the financial implications of the present liability standards as well as any proposed alternatives to them.

Considering that the Superfund program apparently costs the country several billion dollars per year in direct cleanup costs and an unknown amount in legal fees and other transactions costs, we do a very poor job of keeping track of any expenditures related to Superfund other than those made by EPA. This might be acceptable if PRPs and insurers were content with the status quo. But they are not, and one of their complaints is the cost burden associated with the present liability standards. Until and unless PRPs and insurers compile data on Superfund-related expenditures and make this information available to the public, such complaints will be difficult if not impossible to evaluate.

---

[2]Acton and others, *Superfund and Transaction Costs.*

Each of the alternative approaches that we consider in chapter 3 would require major increases in the corporate environmental tax—from a 70 percent increase for one to an 800 percent hike for another. Yet, as we have noted, there is a great deal of uncertainty about each of the building blocks on which our estimates are based: the average cost of a cleanup, the number of sites likely to be affected by each of the options, and the time horizon over which cleanup would take place. Without better information on all these questions, it is virtually impossible to come up with a realistic estimate of the cost of the current program, much less the alternatives to it. Any serious consideration of liability alternatives by Congress will, we hope, be based on more sophisticated analysis of the costs of the current program and the financial impacts of any change in liability.

This report is silent on two important issues relating to changes in Superfund's liability standards and taxing scheme: the effect of any change in liability on the kinds of remedies selected and the ability of EPA to effectively manage a large number of site cleanups at one time. These issues must be borne in mind if serious consideration is given to changing Superfund's liability scheme.

We assume throughout this report that the same remedy would be selected at each NPL site under all five of the options we consider. That is why we increase the Trust Fund by the full amount that PRP spending would be reduced under the various alternatives. The issue is more subtle than this, however. For example, PRP involvement in the site study, remedy selection, and actual cleanup process *may* ensure that the most cost-effective actions are taken at each step of the way. This would be absent if and when responsibility for the cleanup for some fraction of NPL sites reverted solely to EPA and the Trust Fund. Also, to the extent that government spending is seen as "free money," the financing of site cleanups by EPA rather than by PRPs might result in more local pressure for gold-plated remedies. Any serious consideration of a change in Superfund's liability standards should address these potential effects.

A second important issue concerns the number of NPL cleanups that EPA can responsibly manage. We raise this issue most directly in our discussion of Option 5, which would have the cleanup of all 1,115 nonfederal NPL sites revert to EPA. At least some critics have expressed the view that EPA is already doing an unsatisfactory job of cleaning up the relatively small number of sites for which it now has lead responsibility. If this is so, even Options 3 and 4—which would turn over 250 and 580 sites, respectively, to EPA and the Trust Fund—could overwhelm the agency, resulting in a much slower pace of activity or less permanent remedies than we assume, or both. Any significant increase in the proportion of Trust Fund-financed cleanups would have to be preceded by some hard thinking about the scope of EPA's administrative capabilities.

For all the attention paid to the costs of the Superfund program, they may pale in comparison with the costs of cleaning up problems at federal facilities (particularly those of the Department of Energy) and also in comparison with the costs of "corrective action" cleanups that will be required under RCRA.

According to researchers at the University of Tennessee, the Congressional Budget Office, the General Accounting Office, and elsewhere, remediating contamination at the defense weapons facilities of the U.S. Department of Energy (DOE) is likely to be more expensive than the cleanups at nonfederal sites on the NPL.[3] As importantly, depending on the final regulations EPA issues governing remediation of contamination at sites with facilities requiring a RCRA permit (the corrective action program), that cleanup program could dwarf even an expanded Superfund program. Recent estimates by researchers at the University of Tennessee suggest that the total cost of Superfund cleanups is likely to be $151 billion over the next 30 years, while the total cost for DOE facilities is estimated to be $240 billion, and for RCRA corrective action $234 billion.[4]

We do not make these comparisons to suggest that the Superfund program is insignificant or that it should be ignored. Any opportunity to eliminate several billion dollars in needless transactions costs should be seized so long as the downside is seen as being minor, which may or may not be the case here. As we have made clear, we believe this is an ideal time to weigh carefully the strengths and weaknesses of both the current liability provisions of Superfund and possible alternatives to them. Regardless of the outcome of such a review, this nation's approach to site remediation will be stronger for it. We hope these other important cleanup programs come under the same bright light now being directed at Superfund.

[3]Congressional Budget Office, *Federal Liabilities Under Hazardous Waste Laws* (Washington, D.C., CBO, 1990); Milton Russell, E. William Colglazier, and Mary R. English, *Hazardous Waste Remediation: The Task Ahead* (Knoxville, Tenn., University of Tennessee, Waste Management Research and Education Institute, 1991); Testimony of Victor S. Rezendes, U.S. General Accounting Office, before the Department of Energy Defense Nuclear Facilities Panel of the House Committee on Armed Services, March 30, 1992.

[4]Russell, Colglazier, and English, *Hazardous Waste Remediation*, p. 16.

Appendixes

Appendix A

Description of NPL Survey Data

In the first few months of 1992, the Environmental Protection Agency (EPA) sent surveys to all 10 of its regional offices to obtain a variety of site-specific information that Resources for the Future (RFF), the U.S. General Accounting Office, and the Congressional Budget Office had requested on all National Priorities List (NPL) sites. In this report, we include information requested by RFF based on this survey.

EPA collected survey information on 1,003 of the then-1,250 NPL sites. The only sites excluded from this survey were federal facilities and sites where the states had the lead for *all* site activities. In some cases (indicated below) we cross-checked information provided in the survey with site descriptions in EPA's "state books" (*National Priorities List Sites: Alabama-Wyoming*, U.S. EPA, Office of Solid Waste and Emergency Response, September 1990). These books contain brief descriptions of most NPL sites. Where we cross-checked information with the state books, we have so noted. The data used in this report are based on our comparison of the survey responses with the state books.

RFF submitted eight questions to EPA to be included in the survey:

1. Is this an "orphan" site or does this site have "orphan" operable units? (Orphan is considered to be no PRPs have been found or those that have been discovered are not viable.)

2. Is (was) the site owned/operated by a municipality (Y/N)? If available, what percentage of the waste did the municipality contribute?

3. How would you characterize the development of this site? (Illegal dumping activity, owner/operator-only waste contributors, commercial/public solid waste management facility, commercial hazardous waste management facility, etc.)

4. When did the facility stop operating? When did the disposal operation stop? (There may be two dates for this question based on the characteristics of the site.)

5. When was the last date that hazardous waste was accepted at the site/facility?

6. What is the expected TOTAL cost of cleanup for this site/facility? (Please include cost of study, design, cleanup, etc.)

7. If there has been a settlement/order, what percent or number of PRPs settled at this site? What percent of the costs were covered by the settlement?

8. If available, have there been any third-party contribution actions filed by the settling parties against non-settlors (Y/N)?

Responses to some questions (numbers 2 through 5) were much more consistent than responses to others (numbers 1, 6, 7, and 8). We have utilized only information we believe to be reliable. Those data are summarized in the tables that follow.

1. Orphan Site Information

Orphan Site Information (1,003 sites surveyed)

	# of Sites	% of Total
No data	137	13.7
Sites that are orphan sites[a]	104	10.4
Sites that are not orphan sites	762	76.0
TOTAL	1,003	100.1[b]

Note: Thus, for those 866 sites where data were provided, 12 percent (104) were identified as orphan sites.
[a]See table below, "Additional Information on the 104 'Orphan' Sites."
[b]Total not equal to 100.0 due to rounding.

Additional Information on the 104 "Orphan" Sites

	# of Sites	% of Total
Some information on cleanup cost included	70	67.3
Facility closed before 1981	27	26.0
Disposal operation closed before 1981	31	29.8
Last accepted hazardous wastes before 1981	15	14.4
Were owned by a municipality	15	14.4
Were not owned by a municipality	86	82.7

2. Municipal Ownership Information

Municipal Ownership Information (1,003 sites surveyed)

	# of Sites	% of Total
Owned/operated by municipality[a]	174	17.3
Not owned/operated by municipality	725	72.3
No data	104	10.4
TOTAL	1,003	100.0

[a]Of the 174 municipally owned/operated sites: At 83 (47.7%) the facility closed before 1981 and at the same number the waste disposal operation closed before 1981. According to the survey information, 22 of these 174 facilities are still operating.

3. Site Description Information

Site Description Information (1,003 sites surveyed)

	# of Sites	% of Total
No data provided	100	10.0
Data provided	903	90.0

Description of "Owner/Operator" Sites

Of the 903 sites for which descriptive information was provided in the EPA/RFF survey about the origin of site contamination, 261 sites (28.9 percent) are described as "owner/operator-only." We identified an additional 33 sites, based on the survey information, that are not described specifically as "owner/operator-only" but clearly fall into this category. That is, the sole source of contamination at the site was the result of the owner/operator and no other parties. This brings the total number of NPL sites where site contamination was the result of owner/operator activities only to 294, or 32.6 percent of the 903 sites for which information was provided.

To verify the accuracy of these results, we compared the categorizations from the EPA/RFF survey to detailed descriptions of site histories contained in the EPA state books. Of the total of 294 "owner/operator-only" sites identified in the survey, 240 were clearly the result of owner/operator-only activities. At 23 of the sites it appeared, based on the state book descriptions, that the owner/operator was *not* the only source of contamination. At another 23 of the sites we were unable to confirm the owner/operator-only status, and 8 of the sites were not in the state books. Based on this analysis, we refer to 240 NPL sites as being the result of the activities of owner/operators in the text of this report. This is 26.6 percent of those sites for which information was provided. Applying this percent to all 1,115 NPL sites would yield 297 sites in this category.

Description of "Illegal Activity" Sites

From among the 903 sites for which information was provided in the EPA/RFF survey, we identified 127 sites (14.1 percent) that are categorized as resulting from illegal disposal of hazardous substances. To verify the accuracy of these descriptions, we confirmed the survey descriptions with those contained in the EPA's state books. Of the 127 sites, 89 (70.1 percent) were clearly identified in the state books as sites where illegal dumping was the cause of contamination. An example is the New Jersey site, South Jersey Clothing Co., which is described as follows: "operations generated TCE [trichloroethylene]-contaminated wastewaters and sludges that were routinely discharged onto the ground behind the process building...(p. 198, Ibid, New Jersey)." Thus, throughout the report, we refer to 89 sites (just under 10 percent of those for which data were provided) that were the result of illegal disposal. Applying this percentage to the 1,115 sites at the NPL would suggest that approximately 110 sites are the result of this kind of activity.

From information contained in the EPA/RFF survey and confirmed by the state books, the owner/operator was responsible for the contamination/illegal activity at 17 of these 89 sites (19.1 percent). In addition to these 17 sites, another 18 sites that RFF identified as being "owner/operator-only," based on the state books, are identified as sites where contamination resulted from illegal activities. This brings the percentage of illegal disposal sites resulting from owner/operator activities to 39 percent (35 sites).

According to the state books, of the remaining 39 sites, 22 sites were a combination of permitted disposal and illegal dumping, 12 others had unidentifiable sources, and the remaining 4 sites could not be confirmed with state book descriptions.

4. Date NPL Sites "Closed"

This question was asked three different ways (see RFF survey questions 4 and 5 at the beginning of this appendix). The following tables present information received in response.

A. Date Facility Stopped Operating (1,003 sites surveyed)

	# of Sites	% of Total
No data	216	21.5
Sites for which data were provided[a]		
Before 1981	342	34.1
Between 1981 and 1991	248	24.7
Still operating	197	19.6
TOTAL	1,003	99.9[b]

[a]See table A(1), "Additional Information on 787 Sites for Which Data Were Provided."
[b]Total not equal to 100.0 due to rounding.

A (1). Additional Information on 787 Sites for Which Data Were Provided

Date Facility Closed	# of Sites	% of Total
Before 1981	342	43.5
Closed in 1981	34	4.3
Closed in 1982	37	4.7
Closed in 1983	36	4.6
Closed in 1984	33	4.2
Closed in 1985	30	3.8
Closed in 1986	19	2.4
Closed in 1987	12	1.5
Closed in 1988	13	1.7
Closed in 1989	12	1.5
Closed in 1990	9	1.1
Closed in 1991	12	1.5
Closed in 1992	1	0.1
Still operating	197	25.0
TOTAL	787	99.9[a]

[a]Total not equal to 100.0 due to rounding.

B. Date Disposal Operation Closed (1,003 sites surveyed)

	# of Sites	% of Total
No data	295	29.4
Sites for which data were provided[a]		
Before 1981	369	36.8
Between 1981 and 1991	254	25.3
Still operating	85	8.5
TOTAL	1,003	100.0

[a]See table B(1), "Additional Information on the 708 Sites for Which Data Were Provided."

B(1). Additional Information on the 708 Sites for Which Data Were Provided

Date Disposal Operation Closed	# of Sites	% of Total
Before 1981[a]	369	52.1
Closed in 1981	37	5.2
Closed in 1982	33	4.7
Closed in 1983	38	5.4
Closed in 1984	33	4.7
Closed in 1985	29	4.1
Closed in 1986	21	3.0
Closed in 1987	12	1.7
Closed in 1988	16	2.3
Closed in 1989	13	1.8
Closed in 1990	11	1.6
Closed in 1991	10	1.4
Closed in 1992	1	0.0
Still operating	85	12.0
TOTAL	708	100.0

[a]See table B(2), "Additional Information on the 369 Sites Where Disposal Operation Closed Before 1981."

B(2). Additional Information on the 369 Sites Where Disposal Operation Closed Before 1981

	# of Sites	% of Total
Number that were also illegal disposal	54	14.6
Owner/operator-only waste contributor	110	29.8

C. Date Hazardous Waste Last Accepted (1,003 sites surveyed)

	# of Sites	% of Total
No data	587	58.5
Before 1981	263	26.2
Between 1981 and 1991[a]	124	12.4
Still accepting hazardous waste	29	2.9
TOTAL	1,003	100.0

[a]See table C(1), "Additional Information on 124 Sites Where Hazardous Waste Last Accepted Between 1981 and 1991."

C(1). Additional Information on 124 Sites Where Hazardous Waste Last Accepted Between 1981 and 1991

Waste Last Accepted	# of Sites	% of Total
In 1981	19	15.3
In 1982	22	17.7
In 1983	27	21.8
In 1984	11	8.9
In 1985	15	12.1
In 1986	8	6.5
In 1987	4	3.2
In 1988	8	6.5
In 1989	4	3.2
In 1990	1	0.1
In 1991	5	4.0
TOTAL	124	99.3[a]

[a]Total not equal to 100.0 due to rounding.

5. Cost Information

Cost Information (1,003 sites surveyed)

	# of Sites	% of Total
No estimates of cleanup cost provided	504	50.2
Some cost information provided[a]	499	49.7

[a]Although the survey clearly asked for total estimated cleanup cost, our review of the survey forms revealed that for many sites, only partial cost information was received. Based on this *partial* information, average site cost at 499 sites where some information was provided is $27 million.

6. Third-Party Suits

Third-Party Suits (1,003 sites surveyed)

	# of Sites	% of Total
Third-party action taking place	57	5.7
No third-party actions	395	39.4
No data	551	55.0
TOTAL	1,003	100.1[a]

[a]Total not equal to 100.0 due to rounding.

Appendix B

Corporate Environmental Tax Information

Corporate Environmental Tax, 1988
Estimated Revenues by Standard Industrial Classification (SIC)

SIC	Description	Tax Liabilities ($000)	% of Total
	AGRICULTURE, FORESTRY, AND FISHING	854	0.17
	MINING		
10	Metal mining	2,981	0.61
11/12	Coal mining	1,451	0.29
13	Oil and gas extraction	4,548	0.93
14	Nonmetallic minerals (except fuels)	987	0.20
	CONSTRUCTION		
15	General building contractors and operative builders	1,795	0.37
16	Heavy construction contractors	1,357	0.28
17	Special trade contractors	303	0.06
	MANUFACTURING		
20	Food and kindred products	19,153	3.93
21	Tobacco manufacturers	12,156	2.50
22	Textile mill products	2,219	0.45
23	Apparel and other textile products	1,864	0.38
24	Lumber and wood products	4,063	0.83
25	Furniture and fixtures	1,272	0.26
26	Paper and allied products	11,492	2.36
27	Printing and publishing	12,051	2.47
28	Chemicals and allied products	39,225	8.04
29	Petroleum (including integrated) and coal products	37,497	7.69
30	Rubber and miscellaneous plastics products	2,563	0.53
31	Leather and leather products	534	0.11
32	Stone, clay and glass products	3,629	0.74
33	Primary metal industries	9,108	1.87
34	Fabricated metal products	6,011	1.23
35	Machinery, except electrical	26,071	5.34
36	Electrical and electronic equipment	21,277	4.36
37	Motor vehicles and equipment	19,598	4.02
37	Transportation equipment, except motor vehicles	15,485	3.17
38	Instruments and related products	7,703	1.58
39	Miscellaneous manufacturing and manufacturing not allocable	2,793	0.57

(continued)

Corporate Environmental Tax, 1988 (continued)

SIC	Description	Tax Liabilities ($000)	% of Total
	TRANSPORTATION AND PUBLIC UTILITIES		
40-47	Transportation	15,753	3.23
48	Communication	28,730	5.89
49	Electric, gas, and sanitary services	35,214	7.22
	WHOLESALE TRADE	15,690	3.22
	RETAIL TRADE		
52	Building materials, garden supplies, and mobile home dealers	514	0.11
53	General merchandise stores (excludes nonstore retailers)	9,862	2.02
54	Grocery stores, other food stores	2,462	0.50
55	Automotive dealers and service stations	386	0.08
56	Apparel and accessory stores	2,534	0.52
57	Furniture and home furnishings stores	2,514	0.52
58	Eating and drinking places	2,722	0.56
59	Miscellaneous retail stores	3,375	0.69
	FINANCE, INSURANCE, AND REAL ESTATE		
60	Banking	31,754	6.51
61	Credit agencies other than banks	10,565	2.17
62	Security, commodity brokers and services	3,275	0.68
63	Insurance	30,434	6.24
641	Insurance agents, brokers, and service	1,687	0.35
65	Real estate	2,404	0.49
67	Holding and other investment companies	4,575	0.94
	SERVICES		
70	Hotels and other lodging places	1,198	0.25
72	Personal services	539	0.11
73	Business services	4,640	0.95
75-76	Auto repair; miscellaneous repair services	1,181	0.24
781-79	Amusement and recreational services	3,161	0.65
80	Other services	2,751	0.56
TOTAL		487,924	99.31*

*Total not equal to 100.0 due to rounding.

Source: U.S. Treasury Department, Statistics of Income Division, *Source Book 1988*, Corporation Income Tax Returns (Washington, D.C.). Based on a sample of corporate income tax returns used to estimate industrywide financial statistics.

Appendix C

Estimating the Cost of the Five Policy Options

This appendix consists of two sections. Section I presents information on EPA's past and projected expenditures for remedial actions (RAs). Section II presents the assumptions used to calculate the cost to EPA and, by implication, to potentially responsible parties (PRPs) for the 1,115 nonfederal facility sites on the NPL under the current liability standards and the four alternative approaches. We do not calculate the cost for each option of sites to be added to the NPL in the future, nor do we estimate how each alternative approach might affect total cost recovery revenues. Finally, for the sake of simplicity, we assume that the annual RA costs for each of the alternative approaches (Options 2 through 5) will be in addition to EPA's currently planned RA expenditures of approximately $500 million annually.

All information on EPA's expenditures are based on data provided in the FY 1989 and FY 1990 Superfund Reports to Congress. We use FY 1990 information provided by the Office of the Comptroller and the Office of Emergency and Remedial Response of EPA for our baseline assumptions about EPA's *annual* RA and other (non-RA) expenditures. All costs are estimated for the next 10 years, and assume an average site cleanup cost of $40 million.

I. EPA's Expenditures on Remedial Actions

	(in millions)
EPA spending:	
Pre-1990	$1,297.7
FY 1990	309.9
FY 1991	533.6
TOTAL	$2,141.2
Projected post-FY 1991 EPA RA expenditures	5,144.9
Total expected EPA RA expenditures for the current NPL	7,286.1

II. Estimated RA and Annual Costs for Five Policy Options

A. Option 1: Status Quo (the current Superfund program)

	(in millions)
Total estimated remedial action (RA) costs	$44,600.0
Total expected EPA RA expenditures for the current NPL	7,286.1
Implied PRP costs	37,313.9
Implied annual PRP RA expenditures (for each of 10 years)	3,731.4
Estimated EPA yearly RA costs FY 90 and beyond	500.0
EPA other annual program costs	1,033.0
EPA annual expenditures	1,533.9

B. Option 2: Expanded Mixed Funding for Orphan Shares

We assume that there would be orphan shares at 73 percent of the 1,115 NPL sites and that this orphan share would be 10 percent of these RA costs. With total RA costs for 1,115 sites estimated at $44.6 billion, 73 percent of these costs would equal $32.6 billion; a 10 percent orphan share would amount to just under $3.3 billion, or an increase of $325.6 million in Trust Fund expenditures annually. By shifting approximately $3.3 billion in RA costs to the Trust Fund, implied cumulative RA expenditures by PRPs would be reduced to $34.1 billion.

	(in millions)
Estimated total RA costs	$44,600.0
Orphan share (73% of total RA costs x 10%)	3,255.8
Annual increase to Trust Fund (for each of 10 years)	325.6
Annual EPA baseline (including $500 million for RAs)	1,533.9
New annual Trust Fund level	1,859.5

C. Option 3: Liability Release for All Closed Co-disposal Sites

We assume that 250 sites on the current NPL are closed co-disposal sites and that none of these sites is now being cleaned up using Trust Fund monies. Based on previously outlined assumptions, the total remedial action costs for these 250 sites would be $10 billion. Spreading these costs evenly over 10 years implies an annual increase in Trust Fund monies of $1 billion, bringing EPA's annual program costs to just over $2.5 billion. As a result of this change in liability, PRP costs for RAs would be decreased by $10 billion, bringing them from $37.3 to $27.3 billion.

	(in millions)
Estimated total RA costs	$44,600.0
Cost of RAs at 250 sites	10,000.0
Annual increase to Trust Fund (for each of 10 years)	1,000.0
Annual EPA baseline (including $500 million for RAs)	1,533.9
New annual Trust Fund level	2,533.9

D. Option 4: Liability Release for All Pre-1981 Sites

Based on information from the EPA/RFF survey (see Appendix A), we estimate that 580 of the 1,115 sites on the NPL are "pre-1981." The total RA costs for these sites would equal $23.2 billion, which would require an annual increase in the Trust Fund of $2.3 billion assuming these costs are spread over 10 years. Increasing EPA's annual expenditures by $2.3 billion would bring the total EPA yearly costs to approximately $3.9 billion. This shift of RA costs to the Trust Fund would reduce PRP RA costs to approximately $14.0 billion.

	(in millions)
Estimated total RA costs	$44,600.0
Cost of RAs at 580 sites	23,200.0
Annual increase to Trust Fund (for each of 10 years)	2,320.0
Annual EPA baseline (including $500 million for RAs)	1,533.9
New annual Trust Fund level	3,853.9

E. Option 5: Liability Release for Current NPL Sites

To estimate the cost of cleaning up the entire NPL with Trust Fund monies, we simply take the implied PRP RA costs under the current program (see section A above) of $37.3 billion, average them over 10 years, and add them to the baseline yearly costs of EPA's current program. This would require a $3.7 billion increase in Trust Fund revenues annually. PRP RA expenditures for current NPL sites would, of course, be eliminated.

	(in millions)
Estimated total RA costs	$44,600.0
Implied PRP RA costs	37,313.9
Annual increase to Trust Fund (for each of 10 years)	3,731.4
Annual EPA baseline (including $500 million for RAs)	1,533.9
New annual Trust Fund level	5,265.3

Appendix D

Distribution of Trust Fund Revenues for Policy Options

Effect of Increased Corporate Environmental Tax on Relative Share of Annual Total Trust Fund Revenues for Various Funding Sources

Funding Sources	Option 1. Status Quo		Option 2. Orphan Share		Option 3. Co-disposal		Option 4. Pre-1981		Option 5. Current NPL	
	Contribution to Trust Fund ($ million)	Percentage of Total Trust Fund Revenues	Contribution to Trust Fund ($ million)	Percentage of Total Trust Fund Revenues	Contribution to Trust Fund ($ million)	Percentage of Total Trust Fund Revenues	Contribution to Trust Fund ($ million)	Percentage of Total Trust Fund Revenues	Contribution to Trust Fund ($ million)	Percentage of Total Trust Fund Revenues
Petroleum excise taxes	$571.7	37.3	$571.7	30.8	$571.7	22.6	$571.7	14.8	$571.7	10.9
Chemical feedstock taxes	246.1	16.0	246.1	13.2	246.1	9.7	246.1	6.4	246.1	4.7
Miscellaneous income	255.1	16.6	255.1	13.7	255.1	10.1	255.1	6.6	255.1	4.8
Corporate tax, by industry groups										
Chemicals and allied products	37.1	2.4	63.2	3.4	117.4	4.6	223.6	5.8	337.1	6.4
Petroleum and coal products	35.4	2.3	60.5	3.3	112.3	4.4	213.9	5.5	322.4	6.1
Communications	27.1	1.8	46.3	2.5	86.1	3.4	163.8	4.3	246.9	4.7
Electric, gas, and sanitary services	33.3	2.2	56.8	3.1	105.5	4.2	200.8	5.2	302.7	5.7
Banking	30.0	2.0	51.2	2.7	95.1	3.8	181.0	4.7	272.9	5.2
Insurance	28.8	1.9	49.1	2.6	91.2	3.6	173.5	4.5	261.6	5.0
All other	269.3	17.5	459.5	24.7	853.4	33.6	1,624.4	42.2	2,448.8	46.5
Total of corporate environmental tax	461.0	30.1	786.6	42.3	1,461.0	57.6	2,781.0	72.2	4,192.4	79.6
TOTAL	$1,533.9	100.0	$1,859.5	100.0	$2,533.9	100.0	$3,853.9	100.0	$5,265.3	100.0

Note: These estimates are based on FY 1990 Trust Fund revenue figures provided by EPA (Office of the Comptroller, U.S. EPA, Washington, D.C., April 1992). To estimate the percentage of the corporate environmental tax paid by major industries, we applied their percentage share in 1988 (the latest year for which this information is available)—see appendix B—to the total amount of the corporate environmental tax paid in FY 1990.